R0016297136

```
TJ        Billett, Michael.
1075
.B48      Industrial
1979      lubrication
```

Cop.1

	DATE		

Business/Science/Technology
Division

FORM 125 M

The Chicago Public Library

JAN 9 1981

Received

© THE BAKER & TAYLOR CO.

INDUSTRIAL LUBRICATION

A Practical Handbook for
Lubrication and Production Engineers

Other Titles of Interest

DIXON	Fluid Mechanics, Thermodynamics of Turbomachinery, 3rd Edition
DOWSON & HIGGINSON	Elasto-Hydrodynamic Lubrication, SI Edition
DUNN & REAY	Heat Pipes, 2nd Edition
HOLMES	The Characteristics of Mechanical Engineering Systems
O'CALLAGHAN	Energy for Industry
REAY	Industrial Energy Conservation, 2nd Edition
SCHLESINGER	Testing Machine Tools, 8th Edition
SCHOWALTER	Mechanics of Non-Newtonian Fluids
WHITAKER	Fundamental Principles of Heat Transfer Elementary Heat Transfer Analysis

Pergamon Related Journals

International Journal of Heat and Mass Transfer
International Journal of Mechanical Sciences
Mechanics Research Communications
Mechanism and Machine Theory

INDUSTRIAL LUBRICATION

A Practical Handbook for
Lubrication and Production Engineers

by

MICHAEL BILLETT, B.Sc., Ph.D.

Dorset, England

PERGAMON PRESS

OXFORD · NEW YORK · TORONTO · SYDNEY · PARIS · FRANKFURT

U.K.	Pergamon Press Ltd., Headington Hill Hall, Oxford OX3 0BW, England
U.S.A.	Pergamon Press Inc., Maxwell House, Fairview Park, Elmsford, New York 10523, U.S.A.
CANADA	Pergamon of Canada, Suite 104, 150 Consumers Road, Willowdale, Ontario M2J 1P9, Canada
AUSTRALIA	Pergamon Press (Aust.) Pty. Ltd., P.O. Box 544, Potts Point, N.S.W. 2011, Australia
FRANCE	Pergamon Press SARL, 24 rue des Ecoles, 75240 Paris, Cedex 05, France
FEDERAL REPUBLIC OF GERMANY	Pergamon Press GmbH, 6242 Kronberg-Taunus, Pferdstrasse 1, Federal Republic of Germany

Copyright © 1979 M. G. Billett

All Rights Reserved. No part of this publication may be reproduced, stored in a retrieval system or transmitted in any form or by any means: electronic, electrostatic, magnetic tape, mechanical, photocopying, recording or otherwise, without permission in writing from the publishers.

First edition 1979

British Library Cataloguing in Publication Data
Billett, Michael
Industrial lubrication.
1. Lubrication and lubricants
I. Title
621.8'9   TJ1075   79-40526
ISBN 0-08-024232-4

In order to make this volume available as economically and as rapidly as possible the author's typescript has been reproduced in its original form. This method unfortunately has its typographical limitations but it is hoped that they in no way distract the reader.

Printed and bound at William Clowes & Sons Limited
Beccles and London

Contents

List of Illustrations		vi
Preface		viii
Chapter 1	Industrial Lubrication and Production Oils	1
Chapter 2	Lubrication Systems, Synthetic and Solid Lubricants	16
Chapter 3	Circulation Oils	25
Chapter 4	Special Machinery and Material Lubricants	35
Chapter 5	Greases and Temporary Corrosion Preventives	47
Chapter 6	Cutting and Metal Working Fluids	59
Chapter 7	Heat Transfer Applications for Industrial Oils	72
Chapter 8	Industrial Oils in Hostile Environments	86
Chapter 9	Miscellaneous Industrial Oils	92
Chapter 10	Care of Industrial Oils	102
Chapter 11	Industrial Oil Testing	111
Chapter 12	Test Terms and Standard Methods	123
Appendix - Table 1	Viscosity Conversion	131
Bibliography		132
Index		133

List of Illustrations

Fig.	1. 1.	Hydrodynamic lubrication	2
Fig.	1. 2.	Boundary lubrication	3
Fig.	1. 3.	British Standard viscosity grades	5
Fig.	1. 4.	Hydrocarbon types	6
Fig.	1. 5.	Oil oxidation	7
Fig.	1. 6.	Foam and entrained air	8
Fig.	1. 7.	Asperity sulphide formation	9
Fig.	1. 8.	Surface polar films	10
Fig.	1. 9.	Dispersed carbon	11
Fig.	1.10.	Oil in water emulsions	12
Fig.	1.11.	Water in oil emulsions	13
Fig.	1.12.	Fatty acids	14
Fig.	2. 1.	Typical viscosity ranges for industrial oils	17
Fig.	2. 2.	Chemical linkages	19
Fig.	2. 3.	Solid lubricant-graphite	23
Fig.	3. 1.	Types of rolling mill	27
Fig.	3. 2.	Hydraulic principle	28
Fig.	3. 3.	Gear tooth	32
Fig.	4. 1.	Stick-slip behaviour	36
Fig.	4. 2.	Refrigeration circuit	38
Fig.	4. 3.	Texturising process	41
Fig.	4. 4.	Synthetic staple production	43
Fig.	4. 5.	Wire rope cross-section	44
Fig.	5. 1.	Industrial greases - main types	48
Fig.	5. 2.	Grease consistency classification	50
Fig.	5. 3.	Rust formation	52
Fig.	5. 4.	Main classes temporary corrosion preventives	54
Fig.	5. 5.	Dewatering fluid action	54
Fig.	5. 6.	Dipping bath	55
Fig.	5. 7.	Edge coating	56
Fig.	6. 1.	Chip formation	60
Fig.	6. 2.	Cutting fluids - main types	61
Fig.	6. 3.	Tipped cutting tool	64
Fig.	6. 4.	Metal rolling	66
Fig.	6. 5.	Aluminium rolling oil filtration	66
Fig.	6. 6.	Wire drawing	67
Fig.	6. 7.	Forward extrusion	68
Fig.	6. 8.	Backward extrusion	69
Fig.	6. 9.	Deep drawing	70
Fig.	6.10.	Electro-discharge machining	70
Fig.	7. 1.	Heat transfer system	74
Fig.	7. 2.	Laminar and turbulent flow	75

List of Illustrations

Fig. 7. 3.	Steel - time - temperature - transformation curve	79
Fig. 7. 4.	Alloy effect - transformation cooling rate	80
Fig. 7. 5.	Cooling stages	81
Fig. 7. 6.	Cold quench cooling curve	81
Fig. 7. 7.	Hot quenching - martempering	83
Fig. 10. 1.	Oil sampling tube	109
Fig. 11. 1.	Silver ball quench test	115
Fig. 11. 2.	Nickel ball quench test	115
Fig. 11. 3.	Hardenability curve	116
Fig. 11. 4.	Soluble oil corrosion test	118

Preface

The number of different types of lubricants and special oils used in the world of industry is vast. However, despite this great variety, all the more important lubricants, greases and production oils, together with the properties required of them for efficient service in industry, are described in the handbook. In addition to mineral oil based products, the special applications in industry for synthetic and solid lubricants are also discussed, together with the contributions these materials make to increasing possible working temperature ranges.

The book gives practical advice to the working engineer on lubrication under a variety of adverse conditions. It also includes hints on the care, handling and safety aspects of industrial lubricants. The selection and the optimum use of lubricant type and system application are also discussed. The methods of evaluating industrial lubricants, by laboratory and mechanical rig testing, completes the background information given in the handbook.

The subject of industrial lubrication has been treated as far as possible in a practical way rather than a theoretical one, so that the handbook can be of maximum assistance to lubrication and production engineers working in a wide variety of industries. In addition to engineers, the handbook will prove useful to chemists and other personnel connected with production in industry, or studying courses associated with it.

A short bibliography is included at the end of the book to suggest further reading. However, it will be discovered by engineering students that many of the lubrication problems encountered in industry will often have to be solved by decisions based on practical experience, gained by working in a particular industry for a period of time.

The wheels of modern industry are kept moving by the speedy provision of the desired lubricants, at the required quality levels, throughout any country where they may be needed. Much credit for their ready availability is due to the major international oil companies who supply such excellent products and service throughout the world.

Dorchester 1979   M. G. Billett

CHAPTER 1

Industrial Lubrication and Production Oils

The role of the lubrication engineer in practice is now recognised to be one of prime importance in industry. His primary established task is to ensure the smooth running of industrial plant, or production equipment, by efficient lubrication. This achieves savings by the avoidance of costly breakdowns and allows faster rates of production to be obtained. The importance of good lubrication practice is now appreciated to be a means of saving vast sums of money in the total running of a country's industrial activity.

The term industrial oils covers a very wide field of practical applications, as each particular section of industry has demands for oils specifically formulated to its requirements. The specialist products are too numerous to list but the following more widely used oils illustrate the variety of products needed by industry. Steam turbine oils, bearing lubricants, hydraulic oils, gear oils, machine tool lubricants, greases, metal working fluids, quenching oils, heat transfer oils, electrical oils, refrigeration oils, textile oils and rust preventive oils are examples and each individual type has its own quality requirements.

The range of products used may be divided roughly into two categories. Those used in the lubrication of industrial machinery and those used as industrial production oils. An example of the former category is a gear oil which acts purely in the role of a lubricant. An example of the latter is a quenching oil which acts as a production oil by its ability to remove heat quickly when hot steel is quenched in it. This allows the steel hardening process to take place but no lubrication property is required. In other words, the oil has been used to produce a steel item rather than lubricate it.

In certain cases, it is difficult to categorise precisely a product into being solely a lubricant or a production oil. For example, a neat cutting oil used in the metal working industry possesses the dual ability of acting both as a lubricant, by reducing frictional heat between tool and work-piece, and as a production oil by its inherent cooling ability. However, cutting oils are normally classified as production oils.

The term production oil means that no primary lubrication property is needed. By lubrication property is meant the ability of the product to separate physically two surfaces in relative motion to one another; thus avoiding metallic wear by preventing contact between them. This is commonly known as hydrodynamic lubrication (Fig. 1.1). However, this is not always possible as the loading, or dynamic

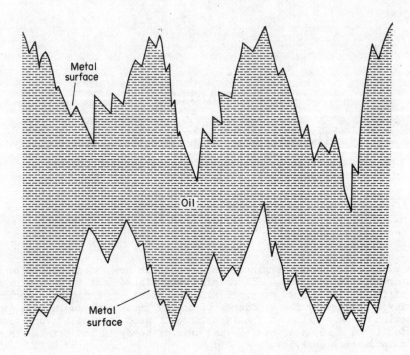

Fig. 1.1. Hydrodynamic Lubrication.

features of the particular machine or mechanism, may cause the lubricant film to break-down. This would lead to wear and surface damage occurring, unless an additional lubrication property was also present in the oil to act under these so called boundary conditions (Fig. 1.2). Industrial machinery lubricants must therefore be suitable, depending upon the particular application, either for hydrodynamic or boundary conditions (or sometimes mixtures of the two).

In the following chapters, reference will be made to turbine oils, hydraulic oils, gear oils, greases, machine tool lubricants and also to production oils for metal cutting and heat treatment of steels. Whilst this might appear to be a random array, it is one which embraces many of the features required of industrial oils. The properties of chemical and thermal stability, friction reduction and load carrying ability (including extreme pressure lubrication) are common requirements of many industrial oils used in different applications.

Before discussing individual oils in more detail, it is perhaps appropriate to consider briefly the recent development trends in industrial lubrication practice. Increasing demands are being made on the oils because industrial progress has made their working environments more severe. This is due to the use of increased loadings, speeds and temperatures which has come about by the development of greater efficiency of industrial design. However, the majority of industrial applications for fluids can still be adequately met by mineral oil (hydrocarbon) based products. In some special applications, it has become necessary to use synthetic (non-hydrocarbon) oils or, in certain cases, solid lubricants. The properties of these types of products will be discussed in detail in chapter 2.

Although synthetic oils are excellent on technical grounds for many applications,

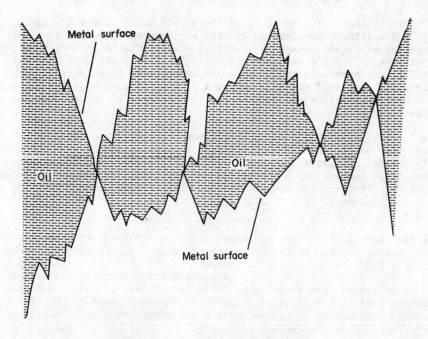

Fig. 1.2. Boundary Lubrication.

and appear to be an obvious choice, they are not always viable on economic grounds. For this reason, industrial equipment has normally to be lubricated with mineral oil based products, especially if a bulk quantity of lubricant is involved in filling a large system.

Types of Industrial Conditions

The general conditions, under which lubricants are desired to work in industry, are extremely varied and often arduous. In a major industry, such as steel making, the lubricated machinery has to function under extremes of heat, dirt and often in the presence of water. In the coal mining industry, the conditions are very dusty and this is exceptionally so for the coal cutting machinery at the coal face. In the paper manufacturing industry, the machinery has to be satisfactorily lubricated often under wet conditions and in the presence of wood chip particles and corrosive chemicals. High temperatures are also used in certain sections of the production schedule and some of the machines are particularly highly loaded. In contrast to this, other machines, such as paper makers, are of extremely sensitive design and require that fine speed adjustments can readily be made to them.

A similar type of delicate control is often required in the textile industry, where a great variety of precision high speed machinery is utilised. In general, it is the usual procedure in the textile industry to employ lubricants in the lower viscosity range. This reduces the total power consumption of the mill, which may be considerable due to the vast number of spindles and other mechanisms to be lubricated. The machinery is normally intricate and often has to work under intermittent conditions. An additional hazard is the possibility of oil contaminating the cloth being processed and for this reason great care must be taken to ensure that the machinery is adequately but not excessively lubricated.

The reverse applies to the brick manufacturing industry, where the emphasis is chiefly on the liberal application of oils and greases to reduce the number of incidences of machinery break-down. This is because the machinery is mainly of a heavy nature and is subjected to rough, dirty outdoor conditions and heavy shock loads. The machinery used in the cement manufacturing industry also works under conditions of heavy loadings and dirt, but some of it also additionally has to function under conditions of heat as, for example, the kilns used for the calcining process.

In direct contrast to this type of environment is that encountered in the food industry. The conditions needed are those of absolute cleanliness and the use of adequate but not excessive lubrication for the food processing machinery. The avoidance of contamination is one of the main features desired and in special applications there is a requirement for medicinal quality oils. The temperatures employed in the industry are usually relatively low, in comparison to most other industries.

Some of the general properties of industrial oils and their additive treatments will now be considered, before discussing the practical applications of the oils in the later chapters. In these later chapters, it will be noticed that some of the oils used, for example, in power generation are encountered in most industries. This also applies to such products as hydraulic fluids, gear oils and greases which are commonly used in a variety of industries. However, some of the oils described in the later chapters will only be applicable to the individual industry they are designed for, such as the special machinery lubricants and the textile fibre lubricants.

General Properties of Industrial Oils and Additive Treatments

The mineral oils employed, for the production of all the various grades of industrial oils, are obtained by fractionation and blending processes. After the lower boiling components have been distilled from the crude oil, the remaining complex mixture of hydrocarbons is subjected to a fractionation carried out under vacuum conditions. This is done to prevent the cracking or decomposition of the higher molecular weight hydrocarbons, which would occur under atmospheric high temperatures needed for distillation. In the vacuum fractionation process, the lubricating oils are separated and collected in fractions of various specified boiling ranges. The separated fractions are refined and may then be blended together to make a long series of viscosity grades for use as industrial mineral oils. The viscosity range will vary between that of a light spindle oil (approximately 10 centistokes at $100^{\circ}F$) and that of a heavy cylinder oil (approximately 1500 centistokes at $100^{\circ}F$).

The use of centistokes as a measuring scale for viscosity is widely accepted internationally. However, individual countries may additionally quote viscosity values on other scales. For instance, Redwood Seconds are popular in the United Kingdom, whilst Saybolt Universal Seconds are frequently utilised in the United States of America. Engler degrees are used as a common viscosity scale in Europe. Viscosity measurements can be converted from one scale to another, provided the values are quoted at the same temperatures. In the United Kingdom, lubricating oils are also classified in viscosity grades rather than in precise viscosity units. For example, there are British Standard Grade designations which quote minimum and maximum viscosity limits (Fig. 1.3). In the United States of America, SAE Numbers (Society of Automotive Engineers) are widely utilised to denote varying viscosity ranges for mineral oils.

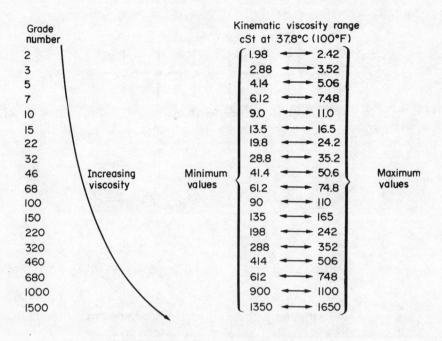

Fig. 1.3. British Standard Viscosity Grades.

Mineral oils are mixtures of vast numbers of hydrocarbons (compounds of carbon and hydrogen), although small amounts of sulphur and traces of nitrogen and oxygen compounds may also be present. The composition of the hydrocarbon mixture depends largely upon which part of the world the crude oil originated. However, most oils are mixtures of paraffins, naphthenes and aromatics (Fig. 1.4). The paraffinic oils are more resistant to oxidation, than the aromatic oils but when oxidation is not a problem, the unsaturated ring type structures of the aromatics allows them to absorb greater quantities of energy before break-down occurs. This specific advantage of aromatic oils is exploited in the nuclear power field when lubricants are subjected to radiation in their working environment. It is also used in the field of high temperature heat transfer where the better thermal stability of the aromatic type oils becomes advantageous. However, when the oxidation stability of the oil is more important, than its thermal stability, for example, in a quenching oil bath, then the paraffinic type oils are preferred to the aromatics. In low temperature applications, such as refrigerator oils, the use of predominantly naphthenic type oils has been traditionally preferred.

In many industrial applications, the selection of a certain hydrocarbon type oil may not be enough to cope with the working conditions imposed upon it. Additives are then incorporated into the oil to enhance its properties for an exceptionally arduous condition. In a limited number of applications, it may be necessary to use a synthetic rather than a hydrocarbon fluid. In certain extreme cases, it may not even be possible to use a fluid and the use of a solid lubricant may be the only alternative. However, in the vast majority of industrial applications the use of mineral oil based additive oils is completely satisfactory.

The additives incorporated into oils to enhance their properties generally go into

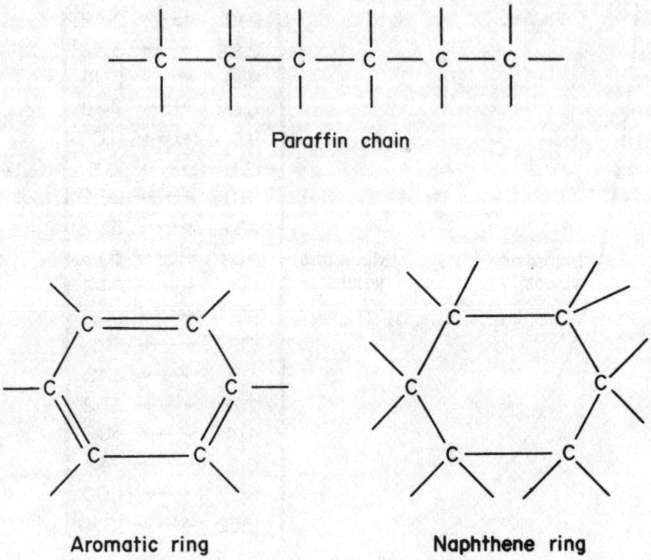

Fig. 1.4. Hydrocarbon Types.

solution with the oil, but sometimes they may be present in colloidal form. The main types of additives used in mineral oils are the oxidation inhibitors, rust and corrosion preventives, anti-foam agents, load carrying and frictional characteristic improvers, pour point depressants and viscosity index improvers.

Oxidation. With regard to oxidation stability, present day refining techniques yield oils of excellent oxidation stability, especially if they are of paraffinic origin. However, in certain applications, the equipment operates under conditions which impose a high demand on the oil; for example, when the oil is in a hot condition in the presence of a copious supply of air and the metallic components of the system. Copper, in particular, acts as a strong catalyst for the oxidative destruction of oil. Oxidation reactions of oils take place in steps, the first products formed being unstable hydroperoxides. These decompose quickly to form various organic compounds such as aldehydes, ketones, alcohols and acids (Fig. 1.5). All these products formed at this stage of the oxidation reaction are usually soluble in the oil. However, in the next stages of the reaction, which consist principally of polymerisation and condensation, insoluble compounds such as gums, resins and sludges may be precipitated from the oil. In practice, the degradation means that an increase in viscosity of the oil will occur and sludge will be deposited. The acidic compounds formed can cause corrosion of the equipment.

The rate at which the oxidative process proceeds depends predominantly upon the quality of the oil. When severe oxidation conditions are present in an industrial application, it is common to use oxidation inhibitors to reinforce the natural inherent stability of the oil. These oxidation inhibitor additives normally function by prolonging the induction period which precedes the main oxidation

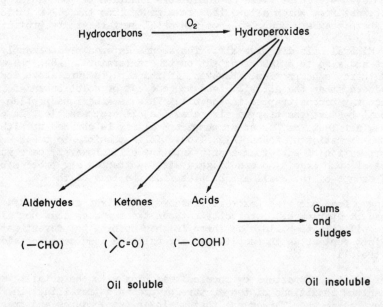

Fig. 1.5. Oil Oxidation.

reaction. The additives may be of the oxidation chain breaker type that interrupt the initial stage of the reaction before it can proceed catastrophically. Alternatively, they may be of the metal deactivator type which minimise the catalytic effect of the metals present in the system by adsorption onto their surfaces, thereby passifying them. In certain applications, it may be necessary to employ both types of oxidation inhibitor in the oil.

Rusting. Oils of good oxidation stability remain in excellent condition for a long time in service, without the oxidative production of lower molecular weight organic acids which cause corrosive acidity. However, it is still possible for rusting to occur in the system because it is a different mechanism from acidic corrosion. Rusting occurs when iron surfaces are in contact with water and air. In practice, it is very difficult to keep small quantities of air and water out of industrial equipment; so rusting is liable to occur irrespective of the condition of the oil. Rust acts as an abrasive and also as a catalyst for oil oxidation. The oil prevents rusting to a certain degree by wetting the metal surfaces, thereby preventing air and water coming into contact with them. The use of rust inhibitors as additives can assist the oil in this respect, by making the oil film become more strongly adsorbed onto the metal surface.

Rust inhibitors are especially useful in large oil circulation systems, where the risk of water contamination is higher. The types of oil used in this application must also have good water separation properties. This is to avoid the possible formation of emulsions which can cause serious trouble in circulation systems, especially if stabilised by the metallic soaps formed as by-products of oil oxidation.

In certain application, it is necessary to incorporate vapour phase corrosion inhibitors into oils to prevent corrosive attack occurring in spaces above the oil level in the system. These types of inhibitors function by possessing relatively high vapour pressures, which allows them to migrate from the oil solution into the air spaces where they are adsorbed onto the metal surfaces to be protected.

Anti-foam. Mineral oils dissolve air. The amount depends predominantly on the air pressure and also to a lesser extent on the temperature. When the air remains in solution there is no problem. However, if the air pressure above the oil is suddenly reduced, then the air will tend to come out of solution and form small bubbles which may become trapped in the oil. This is a serious problem in a hydraulic circuit because the trapped air makes the oil compressible. In other applications, air leaks may occur at pumps or oil may be churned up with air which can cause the formation of foams (Fig. 1.6). It is possible to break such foams by the incorporation of anti-foam agents into the oil. However, care must be taken that the use of such agents does not aggravate the trapped air problem by retarding the rate of escape of the small air bubbles from the body of the oil.

Load carrying. Load carrying (extreme pressure) additives are included in oils when the load, temperature, or velocity between two surfaces does not allow a hydrodynamic oil film to build up. There is then nothing to prevent metal asperities coming into contact, with resulting wear, unless a load carrying additive is present in the oil.

This type of additive functions by chemical reaction with the metallic surfaces but only when the conditions of temperature or pressure prevailing, in the contact zone, are severe enough. This means that at lower temperatures and pressures the additives remain inert. The main chemical elements used for extreme pressure conditions are sulphur, chlorine, phosphorus and lead. They are normally present in the form of oil soluble organic compounds but sometimes sulphur may also be present in its elemental form. The additives are controlled chemical release agents which, on reaction, yield metallic films such as chlorides and sulphides (Fig. 1.7). These films prevent welding and metallic pick-up between the surfaces under heavy duty conditions.

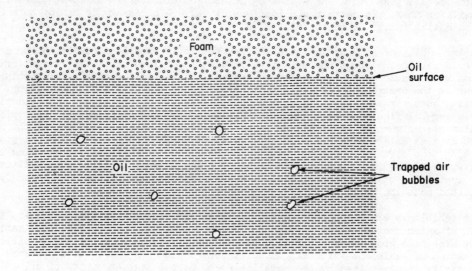

Fig. 1.6. Foam and Entrained Air.

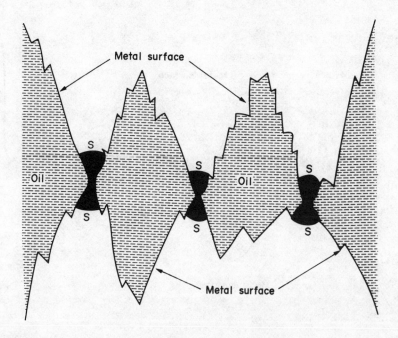

Fig. 1.7. Asperity Sulphide Formation.

Another class of additives is used in oils to modify the friction characteristics. These are the fatty acids and esters. They function by forming a strongly adsorbed polar film on the metallic surface, which reduces the frictional value (Fig. 1.8). The polar type films formed have relatively low melting points compared, for example, to sulphide films. The polar films break-down under extreme pressure conditions and are used as friction reduction agents and not as anti-weld agents.

Pour point. Certain applications for industrial oils demand that they remain fluid at low temperatures. Examples are refrigerator oils and hydraulic oils used in low temperature outside environments. In general, naphthenic oils have lower pour points than the paraffinic oils. The pour point gives an indication of low temperature fluidity. However, pour point depressant additives can be incorporated into paraffinic oils in order to increase their fluidity at low temperatures. The additives are thought to function by inhibiting the honeycombing of the wax separating out from the oil at low temperatures.

Viscosity index. In the majority of applications, the most important characteristic of a lubricating oil is its dynamic viscosity value; i.e. the stress required to shear unit thickness of the oil at unit velocity. This is because under hydrodynamic lubrication conditions, when two moving surfaces are completely separated by an oil film, the only friction source is the oil viscosity. The values of viscosity vary with temperature. However, determinations at various temperatures allow a calculation to be made of the viscosity index. This index can be used to compare different oils, as the higher its value the less is the change in oil viscosity with temperature. The types of hydrocarbons present in the oil and the refining process given to it determine the viscosity index level. For example, paraffinic oils have generally higher index values than naphthenic oils and solvent refined

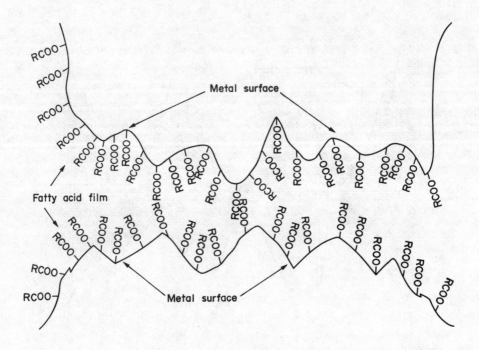

Fig. 1.8. Surface Polar Films.

products have higher values than non-solvent ones. It is possible to increase the viscosity index level of an oil by a very significant amount, if an appreciable degree of solvent refining is carried out.

Solvent refining consists of the mixing and washing of a lubricating oil with solvent. The solvents selected have a predominantly greater solubility for the aromatic type hydrocarbons in the oil, than the other hydrocarbon types which are also present. Eventually, the solvent becomes rich in the aromatics and the resulting solution is then called an aromatic extract. The remaining raffinate, as it is termed, is depleted in aromatics and therefore is richer in the other types of hydrocarbons, such as the paraffins. The solvents are eventually recovered from both the aromatic extract and the raffinate. The solvent refined lubricating oil, obtained from the raffinate, possesses a higher viscosity index value because of its lowered aromatic content.

It is possible to increase further the index level of an oil by the incorporation of a viscosity index improver, which is usually a polymer type additive. It should be noted, that with a non-additive mineral oil above its cloud point, (i.e. the temperature at which wax crystallisation first occurs on cooling) the viscosity is independent of the shear rate. However, when a viscosity index polymer additive is incorporated, this is no longer valid and the viscosity decreases as the rate of shear increases. Under high rates of shear, the polymer is also likely to breakdown with a resulting permanent decrease in viscosity. Careful selection of a stable polymer is therefore of great importance.

<u>Detergent dispersant</u>. In the case of industrial engine oils used to lubricate

stationary diesels, detergent dispersant additives are commonly used for heavy duty applications. These additives prevent deposition and flocculation of carbonaceous materials. These, with other products, are formed during the fuel combustion process and during oxidation reactions. The reaction products are kept in suspension in the oil by the detergent dispersant additives. The additives therefore act as carbon dispersants and they also suppress the symptoms of oil sludging and oxidation.

The fine particles are kept in suspension in the oils because of the surface active nature of the detergent dispersant additive. One part of the additive is preferentially absorbed onto the solid particle surface, whilst the remaining part remains soluble in the hydrocarbon oil phase (Fig. 1.9). The fine particles are stabilised

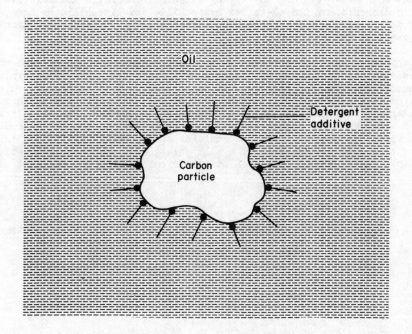

Fig. 1.9. Dispersed Carbon.

and kept bound in suspension by the process. The detergent additives are frequently overbased and made alkaline in nature. The alkali earths, calcium and barium, are frequent constituents of the detergent additives. This alkalinity gives the additives the additional advantage of being able to neutralise any acidic products formed during the combustion process of the fuel in the engine. The sulphur compounds present in diesel engine fuels make the most significant contribution towards the corrosive acids formed.

Emulsifiers. Outside the main field of additive treated industrial oils which are used in the neat oil form, it will be found that in several specialised applications, industrial oils are employed in admixture with water to form stable emulsions. Typical examples are soluble cutting fluids, fire resistant hydraulic fluids and concrete mould oils. However, mineral oil and water are not mutually soluble. A very large quantity of energy has to be expended to shear a mineral oil down to colloidal dimensions, so that it can be dispersed in water to form a stable emulsion.

On an industrial scale, this mechanical method is not usually a practicable proposition, so additives are dissolved in the oil to facilitate the task.

The additives used are called emulsifiers, or surface active agents, and they work chiefly by lowering the interfacial tension between the oil and the water. This allows an emulsion to be readily formed. Afterwards, the surface active agent has the additional task of making the emulsion stable and preventing coalescence back into separate oil and water layers. There are two main types of emulsion, the oil in water and the water in oil. In the former type, the water forms the continuous phase and the oil the dispersed phase (Fig. 1.10). In the latter type, the reverse is true and the oil forms the continuous phase and the water the dispersed phase (Fig. 1.11). Primarily, the type of emulsifier selected will determine the type of emulsion formed. Emulsifiers normally contain components, or groups, which are both soluble in water and oil to varying extents. The ratio and relative influences of these components, or the so-called hydrophilic and lipophilic balance, will normally determine the type of emulsion formed when an emulsifier is present in an oil and water mixture.

The two main types of emulsifier used for the preparation of industrial oil emulsion are petroleum sulphonates and non-ionic surface active agents. The former type are very commonly utilised for the preparation of soluble cutting fluids, which are always used in the form of oil in water emulsions. The petroleum sulphonates are ionic materials, which means they form electrically charged ions in solution. This phenomenon is advantageous when it becomes necessary to dispose of an emulsion after service, because it allows the emulsion to be readily split into separate oil and water phases by the addition of a salt solution or acid. Such materials upset the electric charge stabilisation of the emulsion. This process cannot be done

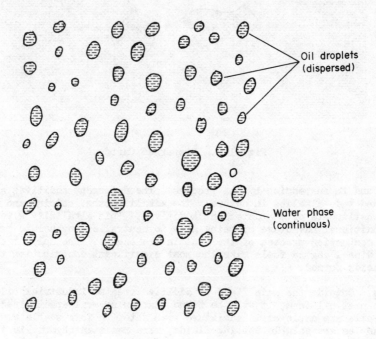

Fig. 1.10. Oil in Water Emulsions.

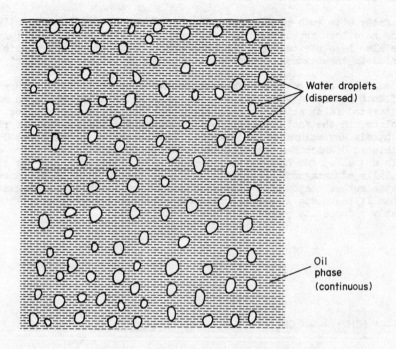

Fig. 1.11. Water in Oil Emulsions.

with an emulsion based on a so-called non-ionic emulsifier and the disposal problem after service is therefore not quite so easy. Special splitting techniques have to be utilised as recommended by the supplier. Alternatively, it may be necessary to employ a waste specialist contractor to collect and dispose of the emulsion. Non-ionic emulsifiers are commonly utilised in the preparation of concrete mould oils and also for fire-resistant hydraulic fluids, when they are of the water in oil emulsion types.

Although non-ionic emulsions are difficult to split, this property means that in service they exhibit inertness and are ideal for use in applications where stability is a prime requisite, for instance, in the presence of chemical contaminants.

Gelling agents. Before leaving the subject of additives in industrial oils, a brief mention must be made of the gelling agents which are sometimes employed as additives. These are mainly applicable to the formation of greases from mineral oil and the specific properties of industrial greases will be discussed in detail in a later chapter. The gelling agents used to manufacture greases are frequently soaps, prepared by reacting alkalis with fatty acids.

Fatty oils. The sources of the fatty acids are frequently vegetable, animal or fish oils. These naturally derived oils provide the vast majority of the fatty compounds used in the general compounding of mineral oils for many industrial purposes. We have already mentioned their use in modifying the frictional characteristics of mineral oils. They are also employed in mineral oils which have to operate in wet environments. The fatty oil, in this case, acts as a surface active agent. It tends to take the water into the body of the oil, in the form of a water in oil emulsion, thus preventing the lubricant film being washed off the surface to be lubricated.

Selected fatty oils such as rape seed, lard, tallow, arachis, sperm, olive, palm and castor have been frequently used for many industrial lubrication purposes. Some have also been utilised for the manufacture of fatty additives which have incorporated in them extreme pressure agents such as sulphur and chlorine.

The various fatty oils possess different compositions (Fig. 1.12). They are a source of both saturated fatty acids, such as palmitic and stearic, in admixture with unsaturated fatty acids, such as oleic and linoleic acids. Castor oil is rather unique, in the fact that it contains an appreciable quantity of ricinoleic acid and hardly any saturated fatty acids. Ricinoleic is an unsaturated hydroxyoleic acid which has practically no action on rubber, unlike the other fatty oils and also, for that matter, mineral oil. This makes the use of castor oil especially advantageous in industrial applications, where it may come into contact with rubber components. However, one of the great disadvantages of castor oil is that it possesses a very high viscosity at low temperatures. This considerably reduces its potential field of activity.

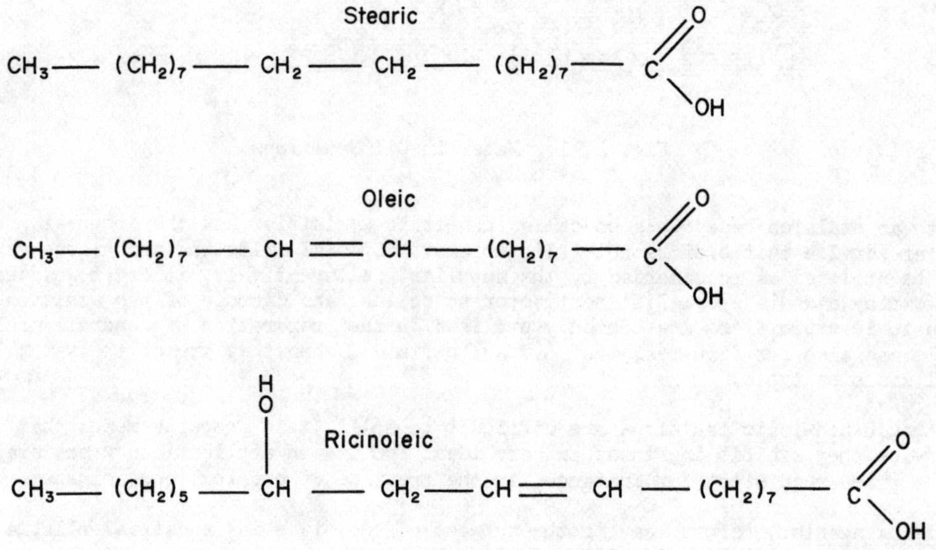

Stearic

$$CH_3 - (CH_2)_7 - CH_2 - CH_2 - (CH_2)_7 - C\begin{smallmatrix}O\\\\OH\end{smallmatrix}$$

Oleic

$$CH_3 - (CH_2)_7 - CH = CH - (CH_2)_7 - C\begin{smallmatrix}O\\\\OH\end{smallmatrix}$$

Ricinoleic

$$CH_3 - (CH_2)_5 - \overset{OH}{CH} - CH_2 - CH = CH - (CH_2)_7 - C\begin{smallmatrix}O\\\\OH\end{smallmatrix}$$

Fig. 1.12. Fatty Acids.

The various fatty oils have different solubility characteristics in mineral oils. Castor oil has only a limited solubility of about 2 per cent. Other fatty oils are much more soluble and are frequently used in the 10 to 20 per cent weight range for the compounding of mineral oils. The solubility is affected by the hydrocarbon types present in the mineral oil and, of course, the temperature.

The main disadvantages of fatty oils, for industrial lubrication, are those of stability and price. They tend to decompose and form gummy deposits at elevated temperatures. They possess, therefore, short working lives in comparison to mineral oils. On lengthy exposure to air at room temperature, there is a tendency for the fatty oils to become sticky and rancid. They are also relatively expensive and many are in short supply. For example, sperm oil supplies have been drastically affected by the international restrictions imposed on the hunting of sperm whales to conserve the species.

However, despite these disadvantages, fatty oils have played and will continue to play an important role in industrial lubrication. This role is not only in the compounding of mineral oil lubricants. In specific applications, fatty oils are utilised in their own rights as lubricants, without admixture with mineral oil. An example is the use of palm oil in the steel industry for the rolling of thin gauge strip. A process for which no mineral oil product can give the same performance. Another example is in the textile industry, which in the past has greatly utilised olive oil and oleines as fibre lubricants. However, this particular application is no longer exclusive to fatty oils, due to their increasing short supply and price level.

CHAPTER 2

Lubrication Systems, Synthetic and Solid Lubricants

LUBRICANT SELECTION AND APPLICATION ARRANGEMENTS

In all the various industries, a prime requisite for the efficient operation of the machinery is for the correct initial selection of a lubricant grade to be made, for each individual mechanism or point to be lubricated. The selection of a grade by a lubrication engineer will often be influenced by the machine manufacturers' recommendations and approvals. This is especially so during machine guarantee periods. However, on other occasions it will be determined by any special environmental conditions the machinery has to operate under. These may even necessitate the use of synthetic or solid lubricants under certain extreme conditions, as will be discussed later in the chapter.

The many different grades of industrial oils find a great variety of applications, in a wide range of machines. However, the general types of lubricating oil involved can be roughly classified as straight mineral oils, detergent oils, hydraulic oils, gear oils, slideway oils and miscellaneous speciality oils. The use of grease is generally more restricted to the lubrication of ball or roller bearings but, on occasions, grease is also used for plain bearings. Grease is selected, in preference to a mineral oil, in applications where the retentive properties of the lubricant are the chief factor. Grease also offers advantages for lubricating mechanisms which are subject to intermittent running. Grease acts as a good sealing material and this property usually makes grease lubrication preferable under dusty conditions, to reduce the risk of abrasive dirt gaining access to the bearing surfaces. The three main general classes of grease encountered in industry are the multi-purpose grades used for the lubrication of roller bearings, the grades utilised for non-retentive applications (such as plain bearings) and the greases employed for speciality purposes. A detailed discussion on individual greases is given in chapter 5.

With regard to the selection of the optimum lubricant application method, hand oiling and greasing will only be an economic proposition when a relatively small number of lubrication points are involved. The high cost of labour favours the use of automatic methods, such as mechanical lubricators, drip feed, ring feed, oil bath and micro-fog systems. Automatic methods reduce the risk of oil starvation, which can sometimes occur when hand oiling methods are used. The distribution of oils and greases in a works by a pipeline system to the individual machines, from a central supply tank, only becomes attractive when the installations are exceptionally large. It is also only a practicable proposition when the layout of the works, or factory, is finalised and no large scale future changes are envisaged.

These circumstances may be found in certain machine shops, where often vast quantities of cutting fluids are utilised in a great number of machine tools. The other main application for centralised systems is found in the steel industry, where many bearings and gears are supplied with lubricating oil by this method, from an underground cellar supply tank. Generally, grease lubricated bearings in the steel industry will also be serviced by a centralised supply system, metering the lubrication points with the correct quantity of grease at the regular and correct intervals required.

The selection of a lubricant grade, for a specific application, will often be influenced by the general state of the machinery and its age and past service record. Without doubt, the most important single physical characteristic of the oil selected will be the viscosity. The precise level chosen will be based on past experience for the particular application (Fig. 2.1). Other characteristics of the oil will also be of importance and these will be discussed in detail in the relevant chapters on the individual oils.

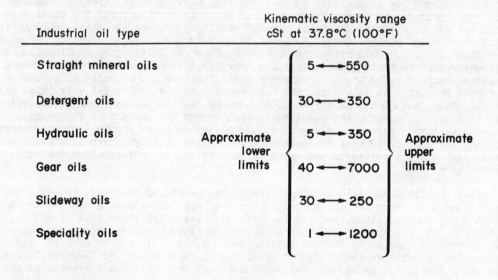

Fig. 2.1. Typical Viscosity Ranges for Industrial Oils.

In carrying out a lubrication survey of a complete works, it is the usual practice to make a list of all the points on all the machines which will require lubrication. It will then be essential to ascertain how frequently it will be necessary to apply lubricants to each of the individual points. This may be required on a daily, weekly, monthly or even longer term basis. Some points may, by design, be lubricated on a total lubricant loss basis, others may use the same lubricant continuously over a long period of time. In the latter case, it will be necessary to drain and refill the system with new lubricant when its estimated useful life has expired. Before a final selection of a lubricant grade is made for a particular lubrication point, it will also be required to consider the method by which the lubricant will be applied to the point. This will probably be by one of the standard lubrication arrangements such as, drip feed, hand oiling or greasing, splash feed, oil cup, ring oiling, mist, force feed or centralised oil lubrication. The system may be one of total loss or may involve complete drain and refill.

Hand oiling methods are used when the relative motion between the two surfaces to be lubricated is slow. The bearing pressure, or the load per unit area, must also be relatively low. However, it is only convenient to use hand oiling, if the oil requirement is infrequent and small. When the motion between the two lightly loaded surfaces continues for longer periods, it is more convenient to use a drip feed oiler. For the application of oil to drip feeds, oil cups, oil holes and similar type lubrication points, it is common practice to use a simple oil can with a squirt attachment.

Splash or drip lubrication methods are sometimes employed as, for example, on certain gears operating at relatively low speeds. For more complicated systems, such as shaft bearings working at higher speeds and loads, the lubrication may be conveniently carried out with the use of a ring oiler. In essence, this consists of a ring which has a greater diameter than the shaft. The ring rotates in contact with the shaft but at a slower speed. The ring dips in an oil trough, picks up oil and transfers it to the shaft and from there to the bearing. The process is carried out on a continuous basis. Other bearings, such as the plain bearings in the spindles of machine tools, for example, are frequently lubricated by pressure or force fed oil. This method is usually employed in lubrication applications in industry when there is a tendency for a heavy load to squeeze the oil out from a bearing surface. The use of pressure fed oil improves the operational performance of such bearings, as it ensures the continuous presence of a relatively thick oil film. Some plain bearings may be lubricated by bearing bushes impregnated with oil. In this type of design, the metal bushes are porous and can absorb for storage approximately 35 per cent volume of oil. The oil is taken slowly from this reservoir during the service life of the bearing. Some plain bearings may not be designed for mineral oil lubrication but may be of the nylon, or polytetrafluoroethylene liner type.

Outside the field of plain bearings, there are the roller bearings. Whereas the plain bearings, such as journal and thrust bearings, consist of a shaft rotating in a bore, bush or liner, the rolling bearings have a different type of design. They consist, for example, of ball or roller bearings, retained by a cage between an inner and outer race. In the rolling bearing, the friction is of a rolling mechanism rather than the sliding type of the plain bearing. Rolling bearings are normally lubricated by grease, because the design will not usually retain an oil in the bearing.

In large lubrication systems, such as mentioned for steel works, centralised methods are utilised for the piped distribution of oils and greases to the points requiring lubrication. Such methods ensure that an adequate supply of lubricant, at the correct pressure, can be delivered to each individual point. Mechanical lubricators are also often fitted to individual machines, possessing a large number of lubrication points, because such an application method is labour saving and convenient. Micro-fog or mist lubrication systems are also encountered in industry, in which the oil is transferred to the lubrication point in the form of a mist mixed with air.

Lubrication Systems, Synthetic and Solid Lubricants

When a particular lubricant grade is selected for an individual application, it will usually be advantageous to determine whether this grade will also be suitable to act as a multi-purpose lubricant for other lubrication points in the factory or works. In this way, it will be possible to optimise and eventually arrive at the essential minimum number of grades needed to lubricate efficiently the entire works.

At this time, a card or computer system sheet can be compiled which will list for each individual machine in the works, the points needing lubrication, their required frequency and the precise lubricant grade to be used. A planned lubrication system has therefore been devised. Operators who carry out the lubrication maintenance can then tick off the lubrication tasks on the card, or sheet, as they are completed. Various codes have been devised to mark clearly the lubrication points on machines. The designated codes will immediately make apparent to the operator the viscosity type, or consistency and grade of lubricant to use. Alternatively, machines may have lubrication reference charts attached to them, instead of the codes. Preventive maintenance can often be carried out and costly breakdowns avoided by the diligence of the operators, who note points where excessive oil consumption may be taking place or where increased noise or vibration is encountered.

SYNTHETIC OILS

The use of synthetic oils, on a sizeable scale, in the industrial field first became fashionable when fire-resistant properties were desired in the lubricant for safety reasons (Fig. 2.2). In the past, such products as chlorinated hydrocarbons and phosphate esters became popular but their use is now severely limited and controlled, due to their possible side effects on the environment. The main concern, especially with the chlorinated compounds, is their long term toxicity and the possible damage they may cause to the environment.

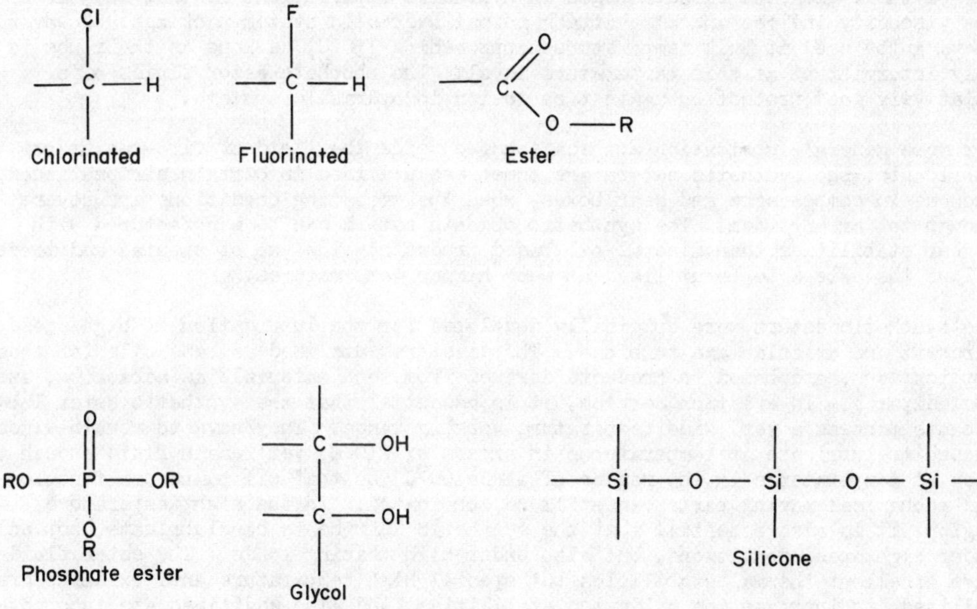

Fig. 2.2. Chemical Linkages.

In the case of chlorinated products, it was found necessary to replace at least 30 per cent of the hydrogen atoms in a hydrocarbon with chlorine, before an acceptable level of fire resistance could be achieved. The chlorinated hydrocarbons were susceptible to hydrolysis and the formation of corrosive acids in service, so care had to be taken during manufacture that they were adequately inhibited against acid formation. The use of other additives to improve their viscosity index levels, and also their oxidation stabilities, made them very useful products in the field of fire-resistant lubricants. However, the present restrictions on their acceptance in industry has now left a void in the provision of an acceptable fire-resistant product.

The perfluorocarbons have an advantage over the chlorinated products, with regard to hydrolytic and overall stability. It would therefore appear they would make excellent fire-resistant lubricants. However, they are not available over such a wide liquid and viscosity range as the chlorinated products, so their use is normally limited to small specialised applications. The contraction in the temperature range, over which the perfluorocarbons are liquid, is due to the effect of the substitution of the fluorine atoms in the hydrocarbons. However, outside the field of completely fluorinated hydrocarbons in the liquid state, it is worth recalling the solid fluorinated polymer, known as polytetrafluoroethylene (PTFE). This polymer is much used for friction reduction in many lubrication applications. It maintains its lubrication properties over a wide temperature range, from approximately minus $50^{\circ}C$ to a region approaching $300^{\circ}C$. Like the other fluorocarbons, it is chemically inert and is both water and hydrocarbon repellent. The great disadvantage of polytetrafluoroethylene is that in the solid form, it possesses a poor dimensional stability. For this reason, it is more often used as an impregnation agent for steel lined bearings and as a coating agent in many friction reduction applications.

To return to the subject of liquid non-inflammable products, the phosphate esters are fairly resistant and stable. They have been appreciably utilised as hydraulic fluids in certain industries, especially in such applications as die-casting machines. One of their greatest disadvantages in hydraulic applications is that they possess low viscosity indices and also attack normal hydraulic system packings. However, they can be used at bulk temperatures approaching $150^{\circ}C$, as long as their use is only intermittent at this temperature level. The phophate ester fluids offer relatively good protection against corrosion in hydraulic systems.

For more general lubrication applications, outside the field of fire-resistant lubricants, the synthetic esters are sometimes utilised in certain circumstances, such as in compressors and gear boxes, when the operating conditions are severe enough to warrant them. The synthetic organic esters can be manufactured with higher stabilities than mineral oil based products. The use of special additives allows the esters to be utilised at much higher temperatures.

The synthetic esters were originally developed for the lubrication of high speed aircraft and aviation gas turbines. The diesters, utilised as base oils for these applications, are based on products derived from such materials as sebacates, azelate and adipates. In aviation service, it is essential that the synthetic ester lubricants possess a very wide temperature working range. They have to remain viscous enough to lubricate at temperatures in excess of $250^{\circ}C$, yet remain fluid enough to flow at temperatures in the region of minus $60^{\circ}C$, so that oil pumps can function and mechanical moving parts can still be lubricated. In the high temperature region, it is also essential that the synthetic lubricants can lubricate, not only under high speed conditions, but also under high bearing loads. The ester fluids have excellent thermal stabilities but special high temperature anti-oxidants are utilised, to increase the oxidation stabilities, and also additives are incorporated for the improvement of the load carrying properties.

Although the diester lubricants were originally designed for aviation purposes, they also offer opportunities to extend the working range of industrial lubricants, when they are substituted for mineral oil based products. However, because of their quality level and price differential from mineral oil lubricants, the diester products are limited to specialist outlets in industry. The type of application, for example, may be a high temperature one which involves the lubrication of the mechanism for life, without the need for lubricant change. The synthetic lubricants have also found several specialist outlets in the automotive field, again under long life, high temperature, conditions.

For less severe operational environments, another class of synthetic lubricant is based on the various derivatives of glycols, known as the polyglycols. They are sometimes utilised, under certain circumstances, for lubrication and also as production oils. The glycol derivatives are available in several viscosity grades. They are sometimes employed as specialist products for the lubrication of synthetic textile fibres and also, when mixed with water, for the heat treatment, or quenching, of steels. One disadvantage of the glycol products, in comparison to mineral oils and synthetic esters, is that they are more prone to oxidation, with the possible formation of corrosive acidic products during service. The polyglycols are also inferior as lubricants and attack rubber packings in systems.

Their great advantage is their miscibility with water. The glycols themselves are not fire-resistant and possess fairly low flash points. They will burn once they become ignited, as will also the synthetic esters and, of course, mineral oils. However, glycols can be readily mixed with water and this property is often taken advantage of to provide a base for the preparation of certain classes of aqueous non-inflammable hydraulic fluids. In the event of a fire, the water has a quench effect and retards the combustion reaction.

For some special applications, another class of synthetic oils, the silicones, are sometimes employed. They are available in a range of viscosities but are extremely expensive. The silicones are chemically inert and are capable of withstanding much higher temperatures than mineral oils, without decomposition. As they are fairly flame resistant, they also have a specialised application in the field of non-inflammable fluids. They are also used in high vacuum technology applications, due to their low vapour pressure characteristics. However, when temperatures are high enough, the silicones will ignite and give rise to vapours containing silicon dioxide.

The silicones are suitable lubricants under a wide range of fluid film conditions. However, they do possess quite low surface tensions which, in fact, are only about two thirds those of mineral oils. The silicones have a tendency to creep rapidly over metallic surfaces and to leak through mechanical seals. This type of property makes silicone fluids ideal as mould release agents and they are often used as such in the processing of plastics. In this application, they are frequently sprayed from an aerosol to minimise the cost and to ensure an even film over the mould surface. The chemical inertness and the colourless nature of the silicones make them ideal for mould release, as the surfaces are not stained, or marked, by the fluid.

Silicones are also used in the form of greases for protecting assemblies, which contain rubber components. Rubbers are prone to attack from mineral oil based products and so conventional greases cannot be used. Silicone greases have both good water resistance and high temperature properties, which makes them ideal for sealing applications. For example, they are frequently applied to electrical equipment and ignition systems to prevent the ingress of moisture.

For rolling bearing applications, various grades of silicone based greases are available to cover a wide variety of bearing speeds and temperatures. The temperature range for the greases can vary from a low minus $40°C$, to a high temperature region approaching 300°C.

Although the silicones are suitable lubricants to use over a wide temperature range, they are inferior lubricants to mineral oils under more normal circumstances. Unlike conventional straight mineral oils, silicones do not exhibit Newtonian properties and when moving slowly they may possess a low viscosity but, when moved more quickly, the viscosity value becomes greater.

The silicones are mainly used in specialised applications, but it is worth also noting that certain silicone polymers are added to many industrial mineral oils, as anti-foam agents, in minute concentrations of a few parts per million. The presence of such traces of silicone has a dramatic effect in accelerating the rate of collapse of a mineral oil foam. The action is possibly caused by the silicone polymer altering the interfacial tension force, existing between the gas and liquid interfaces of the foam.

Before leaving the subject of synthetic lubricants, it should be mentioned that many synthetically produced hydrocarbons find specialised applications in the industrial oil field. Unlike the conventional mineral oil products, which contain a multitude of mixed hydrocarbons, the synthetic hydrocarbons are relatively pure and possess relatively narrow boiling ranges. They may be paraffinic or aromatic in nature. The aromatic synthetic type oils find outlets in such applications as high temperature heat transfer and radiation resistant lubricants. The synthetic paraffinic types may prove useful in the metal rolling field, or in other applications where narrow boiling liquids of good oxidation stability are advantageous.

It is difficult to quote the comparative costs of the various types of synthetic fluids, in comparison to mineral oil. Costs are usually in a state of flux but any future large increases in oil prices may reduce the large price gap, usually found between mineral oil and synthetic grades. The actual cost of the different fluids is also much influenced by the bulk quantities purchased at a time and many other market factors. The presence, or absence, of additives in the fluids will also affect the costs.

In many circumstances, the environmental conditions will dictate the type of fluid for use in a specific application and it will not be a practicable possibility to make a choice between fluids. When a choice is available, factors other than the initial cost should also be considered. For example, it may be thought worthwhile to pay twice the price for a certain synthetic fluid, if it has double the service life of a less stable fluid. Another factor to consider is the relative importance of the lubricant price, in comparison to the total cost of the machinery or process operation if, for example, a costly break-down should occur as a result of using a cheaper fluid.

As a very rough and approximate guide, it is estimated that the cost of a synthetic diester fluid may be at least four times that of a conventional mineral oil based product. Synthetic hydrocarbons will also be several times the price of a conventional mineral oil. The silicone fluids will be extremely expensive, even in comparison to the synthetic diester fluids. The phosphate esters and the chlorinated hydrocarbons (when they could be used) would lie in the price range between the mineral oils and the diester fluids. The polyglycols would be expected to approach the price level of the diester fluids.

SOLID LUBRICANTS

Under certain extreme cases in industry, it may not always be possible to use a liquid lubricant and a solid lubricant may be the only alternative. Solid lubricants are lamellar solids which can be considered to be more readily sheared by an

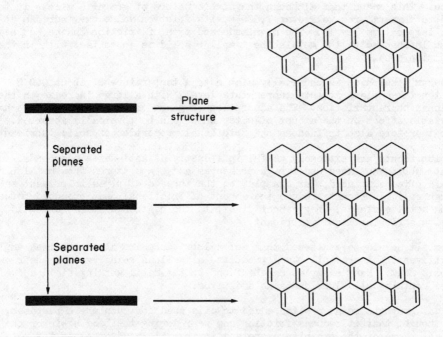

Fig. 2.3. Solid Lubricant-Graphite.

applied force in one specific plane (Fig. 2.3). When a solid lubricant is present on the surfaces of two relatively moving metal surfaces, a gross reduction in the apparent friction is obtained. The most common example of a solid lubricant, found in industry, is graphite. Graphite is available in powder form and it is a soft, light, greasy to the feel, greyish-black crystalline solid. The graphite has a structure consisting of many planes containing a hexagonal arrangement of carbon atoms. There is a relatively large distance between the individual planes. The bonding forces between the planes are weak and the graphite breaks up easily under an applied force into lamillae consisting of thin flakes. For ease of application in industry, various grades of the powder are marketed in the form of colloidal dispersions of the solid, in both water and oil. The colloidal dimensions of the dispersed solid give it a good stability in both these carrier liquids, in which it has no true solubility.

Graphite finds many applications for friction reduction in the metal working industry. When fire resistance is a desired property, for a hot metal process, the colloidal dispersions in water offer obvious advantages. Another commonly utilised solid lubricant is molybdenum disulphide. This is again a lamellar substance and gives friction reduction, due to its preferred slippage in one plane, when under an applied force and separating two relatively moving metal surfaces.

However, with both graphite and molybdenum disulphide, other factors affect the friction characteristics of the lamellar slip and the most important of these are probably the moisture content and the presence of adsorbed gas films on the solid surfaces. In the case of graphite, the friction can increase with a decrease in the moisture content. The reverse applies in the case of molybdenum disulphide and the friction can decrease with a reduction in traces of water on the solid surface. This means that although the stable nature of graphite allows it to be heated to temperatures well over 1,000°C, its friction value may increase at the higher temperature levels. The magnitude and type of friction change, on heating, is much influenced by the metals the lamellar solid is in contact with in the lubrication zone.

Molybdenum disulphide is chemically stable to a temperature of about 900°C. Oxide formation takes place at this temperature level, with a large increase in the frictional characteristics. The adsorption of gases on the solid surfaces has an appreciable effect on the amount of water which can be physically absorbed. This factor therefore also influences the frictional properties of solid lubricants.

Solid lubricants are extremely useful in industry as anti-seize compounds, to protect rubbing surfaces under high pressures and temperatures from metal pick-up. They are often also used, for example, on the threads of pipes to prevent seizure. Although solid lubricants reduce the amount of frictional heat produced between two rubbing metal surfaces, they cannot dissipate the heat once it is formed as can be done with a flowing liquid lubricant.

However, it should be mentioned that solid lubricants are also sometimes employed as additives in mineral oil based products, as well as being used on their own in the solid form. Many examples can be found in the metal working field and also many greases contain solid lubricants as additives.

Colloidal graphite is also often added to oils used for running-in purposes, because it is thought that it reduces friction and wear during the bedding-in of the bearing surfaces. Some of the graphite dispersed in the oil is adsorbed on the metal surface. It then acts as a wetting agent for the oil and thereby speeds up the oil spreading rate, shortening the life of potential local dry spots. Graphite is a good lubricant in its own right. This is a great asset should the situation arise in which the oil supply is temporarily cut off to isolated spots. The graphite will reduce the risk of metal to metal asperity contact occurring, with the possibility of local welding. High temperature spots, caused by excessive rubbing contact may cause break-down of oil in certain cases. The presence of graphite will again be beneficial in this circumstance.

CHAPTER 3

Circulation Oils

Circulation oils in industry are used in enclosed systems and act as lubricants in circulation. They normally are in continuous use and spend little residence time in storage tanks or reservoirs. The type of lubrication involved is in the hydrodynamic region, so the choice of the correct viscosity is of great importance. However, efficient trouble free service also demands good oil stability and resistance to deterioration, brought about by the long and continuous use of the oil.

Steam Turbines

A typical example of a circulation oil is a turbine oil used to lubricate large steam turbine driven electricity generating sets. These turbines possess relatively complicated lubrication systems, which include equipment for forced feed oil circulation and oil operated governors. The oil is present in large quantities in the systems and also in the settling tanks which are used together with other special equipment to maintain the oil in good condition for many years. There are many different pumps included in the large turbine sets and also various pieces of plant such as centrifugal cleaners, which work on the by-pass system. It is essential to maintain the oil free from contaminants and high standards of cleanliness are required during the handling of turbine oils. The oils are usually filtered during the initial filling operation of the turbine.

When in service, the oil is in circulation and its primary tasks are to lubricate and cool the main bearings and governor control gear. The selection of the correct viscosity of the turbine oil grade is an essential requirement so that the oil can carry out these primary tasks. The lubricant must be viscous enough to build up a hydrodynamic lubricating film but at the same time must be thin enough to be able to carry out efficient cooling. The viscosity of the turbine oil must also not vary in service and to minimise any change with temperature, it is essential to use an oil of high viscosity index.

The oil must also have good oxidation resistance in service because any deterioration, with the production of corrosive organic acids and sludge, would considerably shorten the life of the turbine. Turbine oils are expected to have lives of many years, in some cases the same period as the life of the turbine. The use of high quality well refined solvent treated oils, reinforced with oxidation inhibitors, is essential for turbine applications. During service, it is the usual practice to monitor the quality of the turbine oil to ensure that no significant oil deterioration has occurred. This is done by carrying out sampling procedures at regular intervals. The samples are then laboratory tested to determine both the physical characteristics and the existing oxidation state of the oil. Tests may also be carried out to determine the level of oxidation inhibitor present in the used oil, so that an adequate additive level can be maintained during service.

In addition to good oxidation stability, the turbine oil must also be able to separate readily any water which may contaminate it during service in the system. Due to the possible presence of water, the oil must also be able to prevent rusting occurring, both below and above the oil level in the turbine. The formation of rust in a turbine system would dramatically shorten its working life. Turbine oils therefore also reinforced with rust inhibitors to prevent the possibility of rusting occurring.

Another essential requirement of a turbine oil is that it must possess good anti-foam characteristics and it must also be able to release quickly any air trapped in the body of the oil. Anti-foam additives are sometimes included in turbine oils but care has to be taken in their selection so that they do not impede rapid air release from the body of the oil. The requirements for steam turbine oils are therefore very exacting. Most major turbine machinery designers, manufacturers and users detail their own rigid specification for such oils. Only approved oils are therefore normally used in steam turbines.

Roll Neck Bearings

Another example in industry of a circulation oil is that used for the hydrodynamic lubrication of large precision type roll neck bearings, present in multi-stand steel rolling mills. In a simple steel rolling process, the steel is passed through a rotating set of large, smooth, hard steel rolls with a view to reducing its thickness. The mill is called two-high when there are just two rolls (Fig. 3.1). Frequently, extra rolling pressure is achieved by stacking four rolls in pairs, one above the other. The steel sheet is then passed between the two pairs of rollers. This arrangement is called a four-high mill and the outer rolls are larger and press with considerable force on the smaller rolls that actually contact and reduce the thickness of the steel sheet being rolled.

Depending upon the particular rolling process, the steel may be rolled hot or cold. The rollers in the mill are driven by powerful electric motors at a precise speed. There are frequently several individual mills connected in a series, so that the thickness of the steel can be progressively reduced as it passes through the separate mill stands. Such an arrangement is called a multi-stand mill. The thinner the sheet becomes, then the faster the individual rolls must rotate. The steel sheet can be travelling at very high speeds when it leaves the last set of rolls. Due to the considerable amount of heat present in the steel and generated by the rolling reduction process, large quantities of cooling water are used in the rolling operation.

The rollers in the mill are fitted with large precision type roll neck bearings, which have to be lubricated under the prevailing conditions. It is difficult to prevent some cooling water getting into the oil circulation system of the bearings. The amount is often significant despite the efficient design of the roll neck seals. It is therefore essential for the bearing oil to possess good water separation or demulsification characteristics. The bearing manufacturer often lays down his own specification criteria to ensure this characteristic is initially possessed by the oil. In service, if a permanent emulsion is formed due to the water contamination, or if excessive amounts of water are allowed to circulate through the bearings, then considerable damage will occur.

Circulation Oils

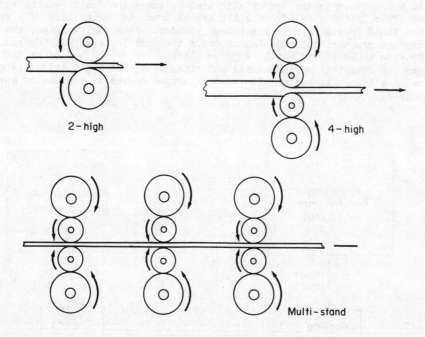

Fig. 3.1. Types of rolling mill.

The correct oil viscosity is also of great importance for the lubrication of the roll neck bearings and high viscosity index values for the oil are usually stipulated. The oil circulation system in steel rolling mills is extremely large and the bearings are on a centralised lubrication system. Metering devices ensure that the correct amounts of lubricant are supplied to the bearings. Pipes and flexible hoses carry the oil as it is pumped to and from the centralised system. The main tanks of the system are usually situated below ground level in a large cellar arrangement. Special filtration and other cleaning devices are incorporated in the system to maintain the oil in a good quality condition. The water content of the oil is monitored on a regular basis to ensure that an excessive amount is never circulated through the bearings.

Hydraulics

Hydraulic oils may also be broadly considered as circulation quality oils, when in hydrostatic systems, as they are also used continuously in an enclosed system with little residence time in a reservoir or storage tank. Hydraulic oils, like turbine oils, must possess good oxidation resistance, good water separation characteristics and good rust prevention properties. They must also possess good anti-foaming characteristics and air release properties. This is of special importance in high pressure systems because the presence of trapped air makes the hydraulic medium compressible.

It is an essential requirement of a hydraulic medium that it remains practically incompressible and also that it remains in a fluid enough condition to allow effective power transmission. In hydrokinetic systems, such as fluid couplings, the power transmission takes place by fluid movement creating kinetic energy. This is in contrast to the hydrostatic system, where the fluid remains virtually static. This latter type of hydraulic system consists essentially of a pump, which supplies fluid from a sump to a working cylinder. From the cylinder, the piston is connected to the particular mechanism where the work is required. Hydraulic systems thus provide a convenient quick response alternative to the use of mechanical linkages, in a variety of industrial and other equipment applications (Fig. 3.2). The use of a hydraulic system also enables a fine degree of control of movement to be obtained.

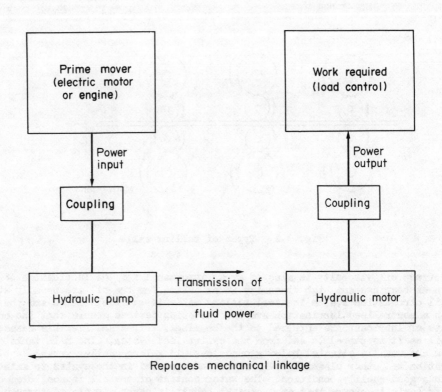

Fig. 3.2. Hydraulic principle.

However, as mentioned earlier, this fine degree of control can only be achieved if the hydraulic fluid remains both incompressible and in a fluid state. Hydrocarbons in the lubricating oil range can, for practicable purposes, be considered incompress However, mineral oils dissolve air and the quantity in solution is mainly dependent upon the air pressure. The temperature of the oil has a relatively minor effect. At normal temperature and pressure, air solubility in mineral oil can be assumed to be approximately 9 per cent by volume. At higher pressures, more will be in solution but as long as the air remains in true solution no problems will occur in the hydraulic circuit. The problem arises when the pressure is suddenly reduced and air bubbles come out of true solution. As indicated earlier, this makes the hydraulic fluid compressible, so it is essential for the fluid to possess good air

Circulation Oils

release and anti-foam properties. The presence of entrained air in a hydraulic fluid leads to lack of control in the operation of the actuating mechanism. It also contributes to hydraulic pump noises.

The other requirement, for efficient power transmission in a hydraulic circuit, is that the medium remains fluid. Mineral oils increase in viscosity, or become thicker, with increasing pressure. However, in the majority of industrial hydraulic circuits, the pressure is usually below 2000 pounds per square inch. Fortunately, mineral oil in this pressure range remains fluid enough for effective power transmission. The approximate relationship of viscosity (η) and pressure (p) can be summarised in the following equation

$$\eta = n \log_e p$$

n is a constant for a particular fluid medium.

Any increase in viscosity of a hydraulic fluid causes an increase in the frictional losses and thereby results in higher system operating temperatures. In practical hydraulic systems, low environmental temperatures usually play a greater role in causing viscosity increases than the pressure increase effect just considered. The pour point of the oil is also of importance under low temperature conditions. It should be well below the lowest operating temperature envisaged for the hydraulic pump. If it is not, then it is likely that the pump may become starved on the suction side with resulting cavitation.

The viscosity of a hydraulic oil is usually selected to satisfy the individual pump used in the system and the valve design. In certain systems, such as the hydraulic control gear of precision machine tools and copying machinery, the fluctuation of viscosity with temperature must be minimised. This can be best achieved by the use of very high viscosity index oils. The same requirement is met in the civil engineering industry when hydraulic equipment is operating in varying outdoor environments. The use of ultra high viscosity index oils in these applications normally entails the use of a viscosity index improver in the oil. As mentioned in chapter 1, these polymer type additives tend to break down under shear with permanent viscosity loss. These conditions are frequently encountered in a hydraulic circuit due to the shearing action of pumps and the fine orifices in valve clearances. It is therefore essential to use shear stable oils in ultra high viscosity index hydraulic applications.

Trends in hydraulic pump design have been towards the use of smaller sizes and higher operating pressures. This development has entailed that the load carrying capacity of the hydraulic oil must meet the increased demands of these pumps, in order to prevent wear. It is usual to include anti-wear agents in such oils, rather than increase the viscosity of the oil to meet this demand. The latter approach would give difficulties with start-up loads at low temperatures. It is essential that the anti-wear agent has good compatibility with all the various metal components present in the hydraulic circuit. It is also important that the anti-wear hydraulic oil can be used in other hydraulic pump circuits without adverse effects.

In certain applications for hydraulic oils, it is advantageous to use a product with fire-resistant properties. Examples are in coal mining and the metal die-casting industries. The rupture of a hydraulic pipe under pressure would give rise to a hydrocarbon spray mist which would be dangerous in an underground mine or in the presence of an ignition source, such as a hot metal in the die-casting industry. In such applications, synthetic or non-hydrocarbon products have been used but now it is quite common to use water based products. Water and mineral oil emulsions and thickened glycol mixtures are used in many hazardous applications where there is a danger of a fire. Although such products will burn if all the water is evaporated from them, the presence of the water has a snuffing action on the flames in the event of a fire.

Compressors

The lubrication of air compressors, such as the larger single and multi-stage reciprocating machines and rotary types, is frequently carried out by a circulatory oil system. A straight mineral oil possessing good stability and demulsification characteristics is generally satisfactory for both cylinders and bearings. However, if there is an appreciable amount of moisture in the air, then it may be necessary to use a separate compounded oil for the cylinders. The build-up of carbon deposits on surfaces sometimes causes fires or explosions. The carbon forming tendencies of the oils are therefore important and are influenced by both the boiling ranges and hydrocarbon types.

With regard to compounded oils, it has been seen that such products consist of mixtures of fatty and mineral oils. The fatty oil may be a blown rape oil or a fish oil, but other vegetable and animal fats are also sometimes employed. The fatty oils act as emulsifying agents when an excessive amount of water is collected in the air compressor cylinders, due to the effect of excessively moist air being compressed. The fatty oil also makes the lubricant film strength greater at the cylinder wall surfaces, as well as incorporating the collected water into the body of the oil in the form of a water in oil emulsion. If a compounded oil was not used under such conditions, then the water would gradually wash off the lubricant film from the cylinder wall. Lubrication failure would then occur. Compounded oils, in general, tend to yield greater amounts of degradation deposits in systems than straight well refined mineral oils.

In the intercoolers of high pressure compressors, the build-up of carbon deposits can sometimes, as previously mentioned, cause fires and explosions. However, this specific type of carbon deposit build up occurs with straight uncompounded compressor oils. As mentioned, it is more a function of the boiling range of the oil used and the hydrocarbon types present in it. It is thought that the carbon deposits, when formed, react with oxygen from the compressed air to form a so-called carbon-oxygen complex. The release of exothermic reaction heat can then act as an ignition source for a potential fire or explosion, if the prevalent conditions are in the critical region. To reduce the risk of fires, it is common practice to use only straight oils of low carbon forming tendencies. This will allow the compressor to operate at higher discharge temperatures for longer periods, without the need for overhaul.

With further regard to fire and explosion hazards, it is extremely important that mineral oil is never used for the cylinder lubrication of oxygen compressors. It will be obvious that the compression of oxygen in the presence of a hydrocarbon will be a very dangerous procedure. In an air compressor the oxygen, of course, is in a minor proportion to the inert nitrogen. When the final compression stage pressure in an air compressor is very high, in the region of thousands of pounds per square inch, it is usually best to use a fairly high viscosity mineral oil to cope with the severe conditions in the cylinders.

Vacuum Pumps

After considering the subject of compressors, it will be worth briefly digressing to discuss the lubricant in vacuum pumps and exhausters. These are used in industry to pump out air from vessels to obtain high vacuum conditions. It will be apparent that an oil of low vapour pressure will be required for the lubrication of the pumps. Normal mineral oils possess relatively appreciable vapour pressures and if used in this application, it would not be possible to obtain a low vacuum due to the high vapour pressure contribution of the oil. It is therefore essential to use a special type of lubricant, prepared so that its most volatile constituents have been removed.

There are two main types of pumps employed to reduce a system down from atmospheric pressure to high vacuum. The first is a mechanical backing pump which reduces the pressure to the range at which the next stage pump, frequently a diffusion pump, can take over. The mechanical pump may be of many types but a common variety is a vane type rotary pump. The oil in the pump assists in obtaining good sealing but there is a limit to the lowest pressure which can be reached in the system. This is due to the leakage of some gas back through the seal and also the higher vapour pressure components of the oil.

Mineral oils are not used in diffusion pumps, which are frequently linked in line with the backing mechanical pump. In this application, silicone oils are frequently utilised because they are available in a range of low vapour pressures. They are, however, very costly but nevertheless essential materials to use in high vacuum technology and other special purpose applications in industry.

Gear Oils

Gear oils may be considered in the general class of circulation oils. Oils are supplied to gears using many various application methods. In circulation systems, the supply may be by forced lubrication or a central oil lubrication system. In the latter method, the oil is delivered to all the parts needing lubrication, independently of the pitch line speeds. The system is designed to ensure that a sufficient volume and pressure of oil is delivered to each individual part. The gear oil is also frequently utilised to lubricate the bearings associated with parts of the gear system. Mist lubrication of gears is sometimes utilised when the speeds are high and prevent the use of the more conventional lubrication method. Oil mist is also directly caused in service by gears operating at high speeds. The churning action of the gears produces the mist.

There are many types of gears used in industry to cover a variety of applications. Generally, the gears are of the involute form; such as spur, helical, double helical, straight spiral bevels and worm gears. The object of using gearing in a machine is to transmit or transfer power. For example, in the main drive of a power unit very large horse powers may be transmitted by the gears. Whilst, in auxiliary systems, gears may be used to transfer power from one section of the system to another. Most industrial gears are made of steel but other metals and combinations are used.

The large number of gear types and applications for them in industry, necessitate a wide range of gear oils being available. The characteristics of the lubricants are designed to meet the conditions and stresses imposed on them in service, so that no gear failures occur due to inadequate lubrication. In order to appreciate these lubrication conditions, it is worthwhile to consider precisely what happens when gear teeth engage and disengage under load.

In the case of spur gears, the load is transmitted between the teeth over a small area of contact, when the gear teeth engage. In theory, the contact zone would be only a line which would transverse the working faces of the engaging gears. In practice, this is not so, due to the elastic deformation of the metal which occurs under load conditions, resulting in a contact area broader than a single line. As soon as tooth engagement commences, some sliding between the two gear surfaces takes place. This sliding component decreases continuously until the pitch line region is reached (Fig. 3.3). At this stage, some rolling also occurs. After the pitch line has been passed, sliding again commences between the disengaging teeth but in the reverse direction.

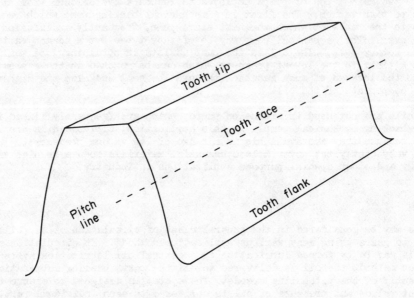

Fig. 3.3. Gear tooth.

These conditions are harsh on any lubricant introduced to prevent metal to metal contact between the gear teeth. The high loads, small contact areas, sliding and rolling forces make it difficult for true hydrodynamic, or thick film lubrication, to prevail. Under certain conditions of loads and speeds it may be possible, but to lubricate effectively under heavy duty conditions, it is usually essential for the oil to have extreme pressure properties. Additive oils are therefore used to cope with these boundary lubrication conditions. The type of gear tooth wear which would occur, if the gears under these conditions were inadequately lubricated, is known as scuffing. The greater the load carrying capacity of the oil, the greater can be the load on the gears before scuffing occurs. This type of scuffing wear is different from the other commonly encountered type, known as pitting.

Pitting is thought to be caused by metal fatigue, which occurs in certain areas of the gears and it may become severe enough to cause tooth breakage. It is difficult to know the precise influence of an oil on pitting. The repeated action of applied stresses at the gear surfaces may lead to metal fatigue, resulting in minute cracks appearing on the gear surface. The oil can then enter these cracks by surface tension effects and the force of the oil pressure. The result may be to enlarge the cracks still further, due to the high pressure of the oil entering them.

Another possible explanation of pitting, is that the repeated stressing can cause minute metal fractures below the gear surfaces, when the tensile strength of the material is exceeded. These fracture zones then progress until the surface is reached and the presence of pitting becomes evident. In the types of possible pitting mechanisms described, it will be concluded that the quality or additive treatment of an oil can do little to alleviate the situation.

There is also another type of pitting, much milder in nature, known as arrested pitting. This probably occurs during the running-in period of new gear sets. High load concentrations may cause this mild form of pitting, which fortunately does not progress during the service use of the gears, after they have been satisfactorily bedded in. There are many other forms of gear tooth damage encountered in service besides scuffing and pitting. The most common are abrasion and scoring.

The design, metallurgy, surface treatment, heat treatment and finishing of the gears play a vital role in their service life. The selection of a gear oil to lubricate them is therefore complementary to their particular history and the conditions imposed on them in service. For many applications, involving only light to medium loaded conditions, straight mineral oils are entirely satisfactory. The most important property is the viscosity of the oil. In the case of bath lubricated enclosed gears, the greater the viscosity the greater is the amount of oil picked up by the gears and therefore the higher is its potential load carrying capacity. Against the use of too high a viscosity is the greater churning friction losses encountered by the gears dipping in the oil. In the case of pressure oil fed gears, the selection of the viscosity level will predominantly depend on the loading to be encountered in the gear.

It has already been seen that oils used in gear lubrication are subjected to thermal stresses. High temperatures are reached in the lubrication contact zone and additional stresses are imposed by the viscous drag of churning and the friction of the bearings in the system. Air is often in contact with the oil, which has as indicated a relatively high bulk temperature. Oxidation stability therefore becomes a problem in service and it is essential for the oil to be resistant in this respect or anti-oxidation additive treated.

In heavy duty applications, such as in steel mill gear boxes subjected to shock loading, it is always essential to use oils containing extreme pressure additives. Traditionally, lead compounds have been the favoured additives used in this application. However, due to environmental considerations, the use of alternative agents such as chlorine, sulphur and phosphorus have been increasingly utilised for this and also other heavy duty gear operations.

In the case of open gears, the main property desired of the lubricant is that it should adhere strongly to the gear surface under all conditions, including wet outdoor conditions. To achieve this aim, it is normal practice to use high viscosity products and in order to obtain the adhesive character required, bitumen based oils are frequently employed. The use of additives and also solvents for ease of application are beneficial.

In the case of worm gears, consisting of bronze and steel wheel combinations, the lubrication requirements are of a different character. The worm gear operates by one surface sliding over the other and therefore it is apparent that friction reduction is of prime importance. It is possible to use straight mineral oils but it is common practice also to incorporate a fatty oil component to ensure the correct frictional characteristics. However, the use of a fatty component lowers the oxidation stability of the oil and therefore it also becomes essential to include an oxidation inhibitor, especially as the temperatures in worm gears can be high. There has been some justification, because of the high temperatures, to use synthetic lubricants in worm gears rather than mineral oil based products. This is especially true in the case of the smaller filled for life units which operate under low speed high torque conditions. It should also be mentioned that pitting appears to be a special problem with worm gears. It is frequently encountered on the bronze wheels but it is doubtful whether it is possible to alleviate this by a particular lubricating fluid selection.

In the case of worm gears operating under both extremely high temperature conditions, up to 200°C, and high pressures, it will be found necessary not only to use a synthetic lubricant but also to have frequent oil change periods. Generally, the main application for worm gears has been in the automotive industry but there are applications in the industrial field, especially for the smaller size units.

In the general field of gear lubrication, the use of a lubricant possessing a low pour point is beneficial. This avoids difficulties in efficient start-up under cold conditions. However, it is not always possible to use an oil in all types of gear boxes. This is because excessive lubricant leakage can occur from certain designs and it is not possible to seal them mechanically against oil loss. In these cases, it is the usual practice to employ a soft gear grease. Care has to be taken, especially under cold start-up or extreme temperature conditions, that the consistency of the grease does not allow "channelling" to take place in the body of the grease. Such a process leads eventually to the gear surfaces running in a dry unlubricated condition, with resulting damage to the gears.

The design of the housing, or casing, is of particular importance in gear grease lubrication. At high speeds, there is a tendency for the grease to be thrown off the gear teeth. It is therefore important that the rejected grease can be diverted back onto the following gear teeth and so enter the meshing zone. At low speeds, there is a tendency for the grease to be slowly squeezed from between the meshing teeth and to collect at the ends of the teeth. Grease lubrication is thus more likely to be efficient with lightly loaded gears. Any grease grade selected must also be suitable for the lubrication of any ball and roller bearings associated with the gears.

The consistency and worked penetration values of greases, classified as soft and also semi-fluid, are given in chapter 5, which deals predominantly with the subject of greases. It will be seen that gear greases, classified as soft, possess a grade number designation of 1. These are normally aluminium soap thickened greases, possessing a smooth texture and good water resistant properties. They have a tacky or sticky consistency, which enables lubricant losses to be reduced when they are used in leaky housings of lightly loaded gears. Gear greases are also advantageous for the lubrication of gears which operate under intermittent service conditions.

For slightly heavier loaded gears, the more recently developed semi-fluid gear greases are suitable for service in spur, bevel and helical gear boxes, especially when they have a tendency to leak with conventional gear oils. These types of gear greases normally possess extreme pressure properties and they lie in the grease penetration range classified as 00, or 000. They are much softer than the soft aluminium soap thickened greases. They also differ from the aluminium soap greases as they do not possess discrete drop point temperatures. This is due to the special types of thickening agents, such as polymers, which are employed to obtain the desired semi-fluid consistencies. They offer, therefore, advantages under higher temperature and high speed conditions. The semi-fluid greases have a smooth texture and, due to their softer nature, they have a much reduced tendency to "channel". The semi-fluid consistency ensures they are less likely to leak from gear boxes than conventional gear oils. The semi-fluid greases are fed to the gears using similar application systems to those employed for gear oils.

A different class of thick gear grease is sometimes employed as an alternative to the bitumen based oils which are usually recommended, as mentioned earlier, for the lubrication of open gears. These types of gear greases are designed for hand application to large slow moving open gears. As with open gear oils, the grease must have extremely good retentive properties and be water resistant, to avoid loss by drip, throw off and atmospheric rain washing, under exposed conditions. The type of gear grease employed in this type of application is normally a calcium soap thickened product, possessing special adhesive characteristics.

CHAPTER 4

Special Machinery and Material Lubricants

INTRODUCTION

Special lubricants are sometimes necessary and used exclusively for the particular machines they are designed to lubricate. Examples in this class are precision machine tool lubricants, rock-drill oils, textile machinery lubricants, refrigerator compressor oils and the synthetic bearing oils used in the roll neck bearings of aluminium cold rolling mills.

Speciality lubricants are also sometimes required for lubrication processes where conventional machines are not involved. These may be called special material lubricants. Examples in this class are textile fibre lubricants and wire rope lubricants. For convenience, the special machinery lubricants will be discussed together first, followed by the special non-machinery or material lubricants in the latter half of the chapter.

SPECIAL MACHINERY LUBRICANTS

Slideways

In machine tools performing precision work, such as profile machining equipment, jig borers and micro-finish equipment, the movement of the machines on their slideways, by necessity, must be slow and precise. These low traverse speeds and the loadings between the surfaces, make it impossible for true hydrodynamic lubrication to take place. The formation and retention of a thick hydrodynamic oil film to separate a moving surface from a stationary one, depends upon the viscosity and density of the oil, together with the velocity of movement. When the velocity is extremely slow, it is difficult to create sufficient pressure within the film to maintain the hydrodynamic condition. When movement ceases, there is an increased tendency for the oil to be squeezed out from between the surfaces. Furthermore, the pressure in the film, under static conditions, is capable of supporting only relatively light loads. Under heavier loads, the two surfaces can no longer be entirely physically separated because the oil film becomes so thin.

The conditions are also not suitable for extreme pressure lubrication. Although extreme pressure properties are desirable in a slideway lubricant to prevent metallic asperity welds between the highly loaded contact surfaces, the use of extreme pressure additives on their own will not give the sliding characteristics required. The conditions on a slideway are unusual in that the loads are high and only very slow speed movement is involved. The main requirement for a satisfactory lubricant to operate, under these conditions, is that it should have some intermediate special friction reduction properties. These are necessary to avoid the stick-slip behaviour frequently encountered when two surfaces move at relatively slow speeds to one another under load. The stick-slip phenomenon prevents smooth movement and gives rise to erratic jerky behaviour (Fig. 4.1). This would be serious in a machine tool required to carry out small precise movements. These are often essential to achieve an exact machining cut or to obtain an extremely accurate surface finish or dimensional tolerance.

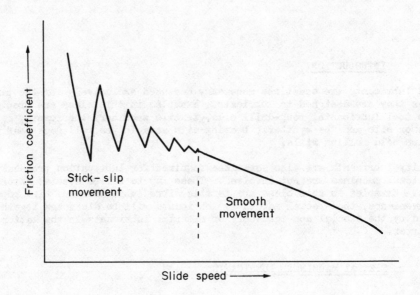

Fig. 4.1. Stick-slip behaviour.

Under stick-slip conditions, a machine tool slide will remain stationary until the applied feed force can overcome the static friction resistance. The slide will then accelerate momentarily but will rapidly slow down and become stationary again. The changes from stick to slip and back again to stick are associated with variations in the friction forces, which prevail under the low speed conditions.

The occurrence and extent of stick-slip will depend greatly upon the design and the material of construction of the slideway. For instance, the use of polytetrafluoroethylene (PTFE) impregnated surfaces can eliminate the occurrence of stick-slip. The low friction characteristics of the slideway created by the PTFE allow steady uniform motion to occur at low speeds. However, the more usual materials for the construction of the sliding surfaces are steel or cast iron and the friction characteristics are then no longer low enough to ensure steady movement.

The basic design of the slideway may also affect the extent of the stick-slip. The various encountered slide types may be flat, v-shaped, dovetailed or perhaps of cylinders and rollers. Combinations of these basic types are sometimes utilised. With regard to the prevention of stick-slip, the ideal combination would be the one giving the lowest friction level. The use of a rolling type slideway would eliminate the occurrence of stick-slip.

However, the utilisation of special coatings and designs are not usually encountered in the majority of machine tools. Other arrangements therefore have to be made to reduce the friction level. Under normal circumstances, the coefficient of static friction is greater than that of kinetic friction; in other words, the force needed to start the movement is greater than that necessary to keep it going. It is this basic difference between the frictional coefficients that leads to stick-slip conditions between two slow moving surfaces. In order to overcome the problem, it has been found necessary to reverse the relative magnitudes of the two friction coefficients. Machine tool slideway lubricants are formulated to achieve this friction change by the incorporation of special additives, which are usually of a polar nature, such as fatty acids. These make the coefficient of static friction become less than the kinetic friction coefficient. The exact ratio of the coefficients required will be somewhat influenced by the materials of construction of the slideway. The slideway lubricant must also have good adhesive properties, so that it will not be squeezed out of the slideway contact zone under load conditions.

Rock Drills

Another example of a special machinery oil is that required for the lubrication of rock drills. These pneumatic tools, especially the larger types used for tunnelling operations, frequently work under wet conditions. It is difficult to keep moisture out of the machine, because it is carried in with the compressed air. It is also common practice to use a stream of water, to assist the drilling operation, for clearing away debris formed at the drill face.

These conditions necessitate that the lubricant used in the rock drill must be able to operate effectively in the presence of water. In order to do this, the lubricant frequently contains fatty materials, or other surface active agents, so that any water can be taken into the body of the oil, by forming an emulsion. The emulsion is of the water in oil type, which means the oil forms the continuous phase and the water the dispersed phase; therefore, the oil can lubricate as it is still in contact with the metal surfaces. Under the prevailing conditions, it is also necessary for the oil to have good adhesive properties; in order to prevent it being washed away by the ingress of water. A further additive needs to be incorporated in the oil to achieve this desired characteristic. It is also essential for the lubricant to have extreme pressure type additives present, due to the shock conditions encountered in rock drilling operations.

Textile Machinery

We will now move on to a third example of a special machinery lubricant. In the lubrication of textile machinery, conventional lubricants are generally used. The viscosities of the oils, especially in spindle lubrication, are kept to the minimum possible, to reduce the power consumption for the mill. This is because vast numbers of spindles are frequently present and wasted frictional energy should be minimised.

However, in a few instances, it is necessary to have special textile machinery lubricants for certain applications. For example, there is sometimes a risk that the cloth or yarn being processed, may become contaminated with the oil used to lubricate the machines. It is possible for oil drops to be flung onto the cloth from rapidly rotating shafts or to drip onto it from overhead mechanisms. The oil stains are a particular nuisance to the industry, because the oil on the fabric also has a tendency to pick up dust from the atmosphere, which further aggravates the soiling problem. There are two main types of special lubricants which can be utilised to alleviate the situation. The first is the use of free scouring textile oils. These oils have emulsifiers incorporated, so that if the oil accidently stains the cloth, it can be washed out in the normal scouring process employed as a finishing treatment in the textile industry. The second method is to use non-throw textile oils. These oils have tackiness additives included in them, to reduce the tendency of the oil to be flung, or to drip, from a lubricated mechanism. This type of oil is very useful in the weaving industry, for example on Jacquard looms, where the moving mechanisms are over the cloth and function by a reciprocating action. Oils are also sometimes used which have both the free scouring and non-throw properties incorporated into the same oil.

Refrigerator Compressors

Our fourth example of a special machinery lubricant is a refrigerator compressor oil. In a typical refrigeration circuit there are three main components, the compressor, the evaporator and the condenser (Fig. 4.2). It is only the compressor section which requires to be lubricated and special oils are needed for this purpose.

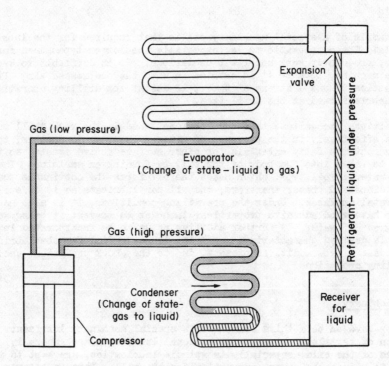

Fig. 4.2. Refrigeration circuit.

There are many different designs of compressors, ranging from the small enclosed rotary or reciprocating types, to the large enclosed vertical reciprocating and the various types of large horizontal machines. However, the exact design generally does not affect the properties required of the cylinder oil, which may be applied either by splash from the crankcase or by direct feed.

It will be apparent that the first requirement of a refrigerator oil is that it should have good low temperature properties. The oil must remain fluid and not separate wax crystals. The pour point of the oil must therefore be low and this is normally the reason why solvent refined naphthenic oils are commonly used. There is also a tendency for the oil to be carried over into the coils of the evaporator and so the low temperature properties of the oil must be good enough to avoid the accumulation of congealed oil on the cold surfaces. Special low wax content oils must be used when evaporator temperatures are very low. The oil must also be completely moisture free, in order to avoid the possibility of ice crystals forming in the refrigeration circuit. Any water present may also react with certain refrigerants, to give rise to acidic components, causing corrosion in the system.

The refrigerants used in large industrial plants are sometimes carbon dioxide and ammonia. Both are inert to lubricating oils. However, in the case of smaller refrigerator units, it is common to use the fluorine compounds or sometimes methyl chloride as the refrigerant. These refrigerants are miscible with lubricating oils and thus have a viscosity decreasing effect on them. The lubricating oil is used in mixture with the refrigerant and the solubility characteristics, over a wide temperature range, are therefore of importance to ensure that the oil can carry out its lubrication function. It should be noted that ammonia is widely used in large installations but carbon dioxide is now more rare, although it was widely utilised in the past. Likewise, methyl chloride is now fairly obsolete in the smaller units and the fluorine compounds are the predominant modern refrigerants.

Sometimes the liquid refrigerant may not only be soluble in the refrigerator oil but also may have a selective solvent action on it. A common example was when liquid sulphur dioxide was employed as a refrigerant. This had a marked tendency to dissolve preferentially aromatic type hydrocarbons from a mineral oil or, in other words, partially to solvent refine it. It was therefore desirable to use a refrigerator oil of low aromatic content in this particular case. However, sulphur dioxide is rarely encountered now as a refrigerant, although it was widely used in the past.

Refrigerator compressors are normally driven by an electric motor as the power source. The lubrication of the motor will depend upon the type and the bearing design employed. Most small sized machines will have plain bearings which will be usually ring oiled, using a straight high quality mineral oil. Large machines may, in addition to oil lubricated bearings, possess ball or roller bearings. These will be grease lubricated and the detailed discussion of greases is the subject in another chapter.

Synthetic Roll Neck Bearing Oils - Aluminium

Our final example of a special machinery lubricant is the oil used in the roll neck bearings of aluminium cold rolling mills. In the cold rolling of aluminium, one of the most important priorities is to avoid staining problems on the aluminium strip being rolled. For this reason, the rolling oil used in the cold rolling process is invariably a fairly volatile narrow boiling hydrocarbon fraction. The use of such a product ensures that any residual rolling oil, left on the finished strip, will be volatile and will readily be removed when the strip is later heated in the annealing furnace. The risk of staining the aluminium strip is thus greatly reduced.

Unfortunately, in rolling mills there is always a risk that the bearing oil used in the lubrication of the roll neck bearings may leak into the actual rolling oil. The bearing oil has a high viscosity and so the situation arises in which the narrow boiling range rolling fluid becomes contaminated with viscous bearing oil. High viscosity mineral bearing oils are not volatile at the annealing temperature and so the aluminium strip will become badly stained. Even minute traces of a mineral bearing oil in a rolling fluid can lead to serious staining problems.

It is not therefore usually possible to use conventional bearing lubricants based on mineral oil. Although great attention may be paid to improving the bearing seals, some leakage appears inevitable. Special bearing lubricants have been developed which are based on synthetic materials rather than mineral oils. The synthetic materials have the high viscosity needed for normal bearing lubrication. However, they are specially formulated to depolymerise, or break-down, into volatile fractions at the temperature level encountered in the annealing process. Any residues can therefore be removed fairly readily from the aluminium strip. The employment of the special bearing lubricants thus helps to avoid the aluminium staining problem, in the event of the bearing oils contaminating the rolling fluid.

SPECIAL MATERIAL LUBRICANTS

Textile Fibre Lubricants

We have mentioned that there are special applications in industry where industrial oils are used to lubricate materials other than conventional machinery. Such a case is the lubrication of textile fibres. This is carried out primarily to reduce the frictional forces encountered as the fibres pass over the various machinery parts used in their processing, thereby reducing the number of fibre breakages which occur. Fibre lubricants also reduce inter-fibre friction and the build-up of electrostatic charges on the fibres which can result in "ballooning" effects, caused by the static repulsion of the fibres.

Fibres can be classified as either natural or synthetic types. Typical examples of the former which require lubrication in processing are wool and jute, whilst examples of the latter are the polyamides and polyesters. In general, most fibres are processed by a similar sequence of operations. Firstly, the individual fibres are prepared, or cleaned, and then made as parallel as possible to one another. The parallel fibres are then twisted, or spun together, to form yarn. Finally, the yarn is made into a fabric by either a weaving or a knitting operation.

In the case of wool there are two distinct branches of the industry, the worsted and the woollen. In the worsted, the longer wool fibres are used. They are made parallel by combing, before being drawn and spun to produce yarn which possesses both fineness and strength. In contrast to this, in the woollen trade the shorter fibres are used but these are not combed and drawn as in the worsted trade. Instead, they are condensed, or entangled with one another, until they possess sufficient strength to be spun. The woollen yarns produced are soft and bulky in comparison to the worsted yarns.

In both worsted and woollen manufacturing industries the wool has to be initially scoured to remove dirt, before processing can be carried out. The scouring process removes much of the natural wool grease lubricant present on the fibres and therefore before further processing can be carried out, it is necessary to relubricate the fibres. Vegetable oils have been much used for this purpose. In the worsted trade, neat olive oil has been traditionally employed but mineral oils blended with fatty oils, are now finding acceptance as combing oils. In the woollen trade, the oleines (impure forms of oleic acid) were traditionally used. Mineral oils with fatty oils are now frequently encountered and it is also common practice to use emulsifiable

Special Machinery and Material Lubricants   41

mineral oils. These are frequently mixed with water and applied in emulsion form to the raw wool.

Another sector of the woollen industry is the "shoddy" trade. This utilises rags which are disentangled to yield the short fibres. These are then processed as in the woollen industry and emulsified oils in water are used for the lubrication of the fibres.

Similar type emulsion oils are used in the processing of jute. The emulsions are often stabilised by the incorporation of an emulsifier in the oil. However, sometimes the emulsions may be formed mechanically, by finely dispersing the oil in the water by a vibrating reed without the need for an emulsifier. In the case of jute, the emulsion softens the woody parts of the material, as well as assisting the further processing by lubrication. It should be mentioned that the use of oils and emulsions in the treatment of natural fibres, also greatly reduces the amount of dust formed, by vibration and shaking of the fibres, in the initial stages of their processing. The reduction of the dust load in a mill makes the atmosphere less of an health hazard.

In the field of synthetic fibres, the types of lubricants used are designed for the particular process being carried out. Synthetic fibres have no natural lubricants present and any required for processing therefore have to be added. As an example, we will consider the process known as "texturising", because there has been an increasing tendency to carry out the process of "texturising" or "bulking" synthetic fibres. Synthetics are normally in the form of straight fibres and therefore they lack the bulkiness or weight of a natural fibre such as wool, which has a hairy type structure. Texturising processes change the geometry of the synthetic fibre to yield one with a zig-zag configuration, which thus has a greater weight per unit length than the original straight fibre (Fig. 4.3). Normally, texturising involves the heating of the fibre as an essential part of the process. The effect of this

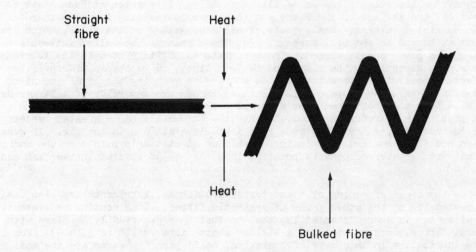

Fig. 4.3. Texturising process.

heat is also to destroy the original "spin-finish" put on the fibre surface by the manufacturer, both as a lubricant and as an anti-static.

It is therefore necessary to relubricate the fibre after the texturising process has been completed. The role of the lubricant is primarily to reduce friction and thus fibre breakages in the further processing operations. The lubricant used is called a coning oil because it is applied to the texturised fibre before it is wound into a cone shaped package on a coning machine. The lubricant is applied by a roller-trough arrangement. The fibre rubs against an oiled roller and the roller and fibre speeds are adjusted to pick-up a precise amount of lubricant on the fibre. This is usually in the region of 2 to 4 per cent by weight.

The oiled cone shaped package is then usually sent to the knitter for processing. Large quantities of bulked fibres are used in the knitting industry and the oil on the synthetic texturised fibre is of benefit to the knitter, as a fibre lubricant in the knitting operation. The oil normally contains a surface active agent or emulsifi so that the oil can readily be scoured from the yarn when required. It is important that any residues left must not interfere with subsequent dyeing operations. The presence of the surface active agent in the lubricant has a secondary effect in that it is able to pick-up traces of moisture from the atmosphere. This has the effect of reducing the electrical resistivity of the synthetic fibre surface during processing and thereby prevents the build-up of electrostatic charges.

As an alternative to a mineral oil based coning oil, containing a surface active agent, the employment of a synthetic oil has some merits. For example, the use of a glycol based product has an immediate obvious advantage, in that it is completely soluble with water. This means it can be readily removed from the fibre when require The danger of residues left on the fibre interfering with the dyeing process is thus avoided. Glycol based products act as effective fibre lubricants and anti-static agents. Their great disadvantage is that they tend to become corrosive towards textile processing machinery, when they gradually become oxidised during use. This has greatly impeded their general acceptance in the industry.

With certain types of coning and high speed winding machines, there is sometimes a tendency for the oiled fibre to oscillate rapidly. This movement causes oil to be thrown off the surface of the fibre and therefore contaminate the atmosphere of the works by mist formation. Small pools of oil also gather on the floor, which present a further hazard to the mill workers. For these reasons, so-called anti-splash coning oils have been developed. These contain an additive to make the lubricant adhere more strongly to the fibre and be less likely to be thrown off the fibre surface. These oils have met with some success in reducing the total amount of oil thrown. However, the drops of oil which are thrown are generally of a larger droplet size than those obtained from a normal non-additive type coning oil. The oil mist problem is therefore reduced, because the thrown oil has a greater tendency to fall to the floor, rather than remain in a stable mist form in the air. In general, design modifications and the placing of shields at strategic points on the machines has met with greater success in practice than the use of special anti-splash coning oils.

Outside the field of coning oils and texturised fibres, lubricants are also used in the processing of the other forms of synthetic fibres. Long continuous synthetic fib are often cut in short "staple" lengths, so that they can readily be mixed with natural fibres, such as wool, of a similar short fibre length (Fig. 4.4). For the fibre lubrication in this type of operation, solid paraffin waxes are the most frequently employed. The waxes are often utilised in the form of discs, which rub against the fibre being processed.

When the synthetic fibres are employed in their long continuous fibre form, or uncut state, it is no longer possible to use a solid paraffin wax. The continuous fibre

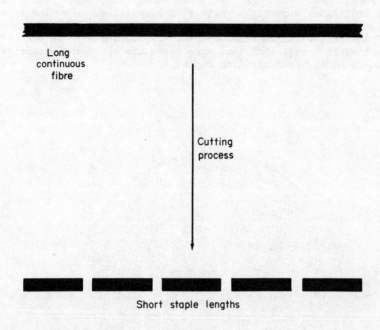

Fig. 4.4. Synthetic staple production.

has a tendency to be sharp and to cut into the wax. Therefore, in this case, it is necessary to employ liquid lubricants. It is apparent that the lubrication requirements of synthetic fibres will be appreciably affected by the physical form they possess during processing. Examples have been given for texturised, continuous and staple forms. The chemical nature of the fibre also affects the lubrication requirements. Two of the main classes of synthetics are the polyamides and polyesters. These different chemical types vary in their frictional characteristics and also their retentive powers for lubricant residues. Other synthetics, such as the acetate fibres, will again possess different surface chemistry characteristics.

<u>Wire Rope Lubricants</u>

Wire rope lubricants are further examples of products used to carry out the lubrication of materials other than conventional machines. In addition, the wire rope lubricants are required to act as corrosion preventives in this particular application.

Wire ropes have many uses in industry. Haulage applications exist for collieries, cranes, overhead ropeways and lifts. Wire ropes are also used for non-haulage purposes, for example, as support ropes for bridges and other engineering projects. In this type of service, they are known as standing ropes.

Although wire ropes are not classified as conventional machines they can, in practice, be viewed as complicated pieces of machinery. They are frequently constructed by

the winding of the steel wires around a central core made, for example, of hemp
(Fig. 4.5). In service, the surrounding wires are subjected to stresses and strains
of various magnitudes. When the rope bends under load, there is rubbing contact
between the individual wires in a variety of geometries. When the rope passes over
a pulley block or drum, there is additionally the rubbing contact between the external
surfaces of the rope wires and the object it is running with in contact. To avoid
wear, it is therefore essential that wire ropes are efficiently lubricated, this
applies both to haulage and standing ropes.

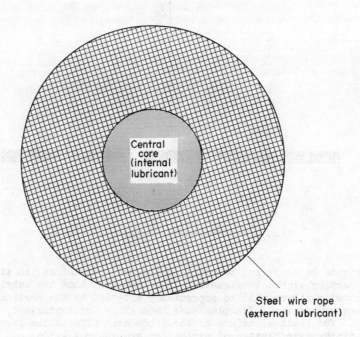

Fig. 4.5. Wire rope cross-section.

Wire ropes are manufactured in a great variety of circumferences, lengths and strength
to satisfy the many design demands of industry and civil engineering. To illustrate
the extremes, a very long rope was once manufactured in the United Kingdom which
measured nearly nine miles long. It had an approximate circumference of three inches
and weighed nearly twenty nine tons. An even heavier rope of approximately eighty
one tons weight has also been manufactured in the United Kingdom. This rope was
about four miles long but had a circumference of seven inches.

Certain ropeways in industry have included as integral parts of them such items as
carriage running wheels and guide sheaves. These will also have to be lubricated,
in addition to the wire rope itself. The bearings used in the sheaves and wheels
may be either of the plain or ball bearing type. The former type may be lubricated
with oil, often the same as used on the exterior rope. The ball bearings will be
grease lubricated.

Special Machinery and Material Lubricants

The lubricants used on the wire ropes themselves vary in type. Sometimes the working environment of the rope plays a role in determining the type of product to use. For example, when wire ropes are used in mining applications, it is essential that any lubricant employed must be of a high flash and fire point to reduce the potential fire risk. In open air applications for standing ropes, this requirement is obviously not so critical. In mining applications, it is also essential to employ products which will not give rise to toxic gases or vapours, in the event of a fire or under high temperature conditions. At one time, there was thought to be a good case to use synthetic fire-resistant wire rope lubricants for mining operations. However, the types of products which could economically be selected were likely to breakdown under high temperature conditions to yield products which would probably have been toxic on inhalation. The use of chlorinated products in this application, for example, was also liable to cause rope corrosion due to their possible hydrolytic instability. However, in mining operations and in all general rope maintenance it is essential to use a lubricant even though the product may be combustible.

The type of lubrication carried out can be classified as either internal or external. During the rope manufacturing process, it is frequent practice to impregnate the core with a lubricant. In service, this acts as a reservoir for the internal lubrication of the rope. The type of lubricant used is usually of a fairly viscous nature, so that it does not drain away too quickly from the centre of the rope. On the other hand, it must have enough fluidity to permeate through the internal wire rope mesh, in order that it may function effectively as a lubricant. The petroleum jellies, wax oil and bitumen oil mixtures are typical products used for this application.

The exterior of the rope must also be lubricated in service, to reduce friction and wear. The product used for the exterior lubrication must also act as a corrosion preventive to inhibit atmospheric attack on the steel wires. The exterior lubricant is subjected to considerable mechanical work and to conditions which tend to wipe the lubricant off the surface of the rope. It is obviously apparent that the lubricant used must have excellent adhesive properties. It also frequently encounters a wet and dirty environment which presents an additional hazard. Rain, for instance, tends to wash off the lubricant whilst pollutants in the atmosphere cause corrosion of the rope.

Lubricants may be applied for exterior lubrication either by hand or by automatic lubricators. If it is done by hand, the product may be a water repellent grease or a viscous bitumen oil mixture. In order to assist the application by hand and to avoid the necessity to heat the product, it is common practice to use solvents to thin down the bitumen blend. This enables better penetration of the mixture into the wire rope to be achieved, but time must be allowed for the solvent to evaporate after application. This is so that the residual material left on the rope can regain the viscous nature it possessed before the solvent addition. If inflammable solvents are used, then appropriate care must be taken to avoid any fire risk.

In the case of exterior application by automatic lubricators, solvent diluted grades are not used. The viscosities of the bitumen and oil mixtures which are frequently utilised must be suitable for the type of lubricators employed.

Wire ropes may be exported to many parts of the world and they are used in a variety of climatic environments. The lubricants must be able to protect the rope from corrosion under all the conditions met in transportation and service. In addition to the standard ropes, manufacturers make many special ropes for specific applications. The types of lubricant used are usually then detailed and specified by the manufacturer.

During the service life of a rope, it is impossible to know the exact internal condition at any particular time, because it can never be visually examined. Manufacturers will normally advise on the expected working life of a rope in a specific application. This life estimate will be based on their past experience and will include a considerable safety factor. This is essential, as a sudden unexpected rope breakage in a mining, steel or engineering application could have a serious or even disastrous consequence. This may not only be in material damage but may also endanger human life. The correct and regular lubrication of wire ropes will reduce the risk of unexpected breakage due to wear or corrosion.

CHAPTER 5

Greases and Temporary Corrosion Preventives

GREASES

Greases are used in many industries. They form a different class of lubricant to those described in previous chapters, which were all fluid oils. Greases are solid or semi-solid products and are used in applications where a fluid oil would run out of the lubrication zone. The retentive properties of greases make them especially suitable for the lubrication of ball or roller bearings and in certain instances, plain bearings. The bearings may either be pre-packed with grease on assembly, or provided with a grease nipple, or cup, so that grease can be applied when desired.

In general terms, greases may be viewed as dispersions of thickening agents in lubricating oils. The usual thickening agent is a soap, which is a product formed by the saponification of a fatty acid with an alkali. Various types of soaps exist and are used to give different properties to the greases prepared from them (Fig. 5.1). Calcium soap based greases have good water resistant properties but have the disadvantage of an upper temperature limit of use of approximately $50°C$. Sodium based greases, on the other hand, have poor water resistance but can be used at higher temperatures, normally in the region of $90°C$. Lithium soap greases possess both advantages. They are water resistant and can be used at temperatures up to approximately $120°C$. This makes them useful in a wide range of applications and for this reason they are called multi-purpose greases. The vast majority of bearings used in industry have running temperatures well within the capability of the lithium soap thickened greases.

Other speciality greases can be made by the use of other soaps. For example, aluminium soaps give rise to greases possessing particularly strong adhesive properties, which minimise leakage from bearings. Complex or mixed soap greases are an example of another type of soap grease. Sometimes, for special applications, fluids other than mineral oils are used. The silicone greases lubricate bearings which have to operate under extremes of heat and cold.

It is possible to use thickening agents other than soaps to prepare greases. Examples of the non-soap type are the various clay and polymer thickened greases. The clay thickened greases are very useful in special high temperature, low speed applications; such as the lubrication of oven chains and kiln car wheel bearings. However, the presence of the mineral oil limits their upper temperature use to approximately

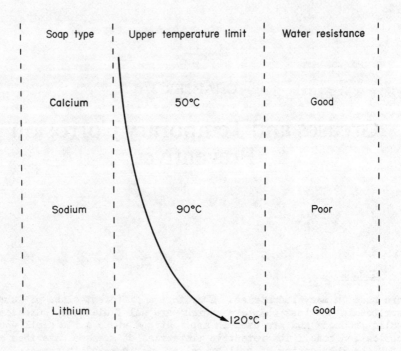

Fig. 5.1. Industrial greases - main types.

200°C. The polymer thickened greases are used when a semi-fluid type grease is required, for example in certain gear box applications where leakage is a problem.

A physical examination of a soap thickened grease will show it to be of a fibrous nature. The fibre structure, responsible for thickening the oil to form a grease, will depend predominantly upon the type of soap used and the method of manufacture. The fibre length may be short which results in a smooth buttery texture. This type of structure is particularly advantageous in a bearing application where low viscous drag is required. Alternatively, the fibre length may be long which will result in a stringy type texture. This type would be beneficial in applications where high speeds are used, which tend to encourage grease throw out by centrifugal action.

The two main components of a grease, soap and oil, both possess excellent inherent lubricating properties. However, the type of structure present may also influence the lubrication efficiency of a grease in a bearing application. For instance, if excessive oil can exude from the structure, then leakage may occur from the bearing through the seals. On the other hand, if the type of structure allows too little oil exudation, then the lubrication mechanism may be detrimentally affected.

A factor which greatly influences grease structure is the amount of shear it undergoes in a bearing. The effect of shear, or working of the grease, is to cause break-down of the structure with the result that a stiff grease may assume the properties of a lower consistency grease. Greases should therefore be of good mechanical stability.

Grease structure is important not only in bearing applications but also in obtaining good pumpability characteristics for the grease. Pumpability is of significance, because greases are often supplied to the various lubrication points of machines by a centralised lubrication system. The grease is pumped, in metered amounts, through pipes from a central supply tank. The grease has sometimes to travel along relatively long pipe lengths. In order to keep the pumping pressures within acceptable limits, the grease must have good flow characteristics.

Another related phenomenon is the "slumpability" of the grease. This is important, for instance, if grease is pumped from the bottom of a supply tank through a pipe. When the grease in the region of the pipe entrance has been pumped away, it is essential for further grease to take its place continuously for the pumping to be able to continue. The grease must therefore be able to fall or "slump" readily from the sides of the tank to the pumping zone.

The temperature of use of the grease also grossly affects the structure. If the temperature is high enough, the grease will undergo a transition from the semi-solid to the liquid state, with complete structure break-down. It is therefore essential that soap thickened greases are used at temperatures well below their so-called drop points.

For high temperature conditions, the oxidation stability of a grease is of special importance and oxidation inhibitors are frequently employed. The type and quality of the base oil present in the grease determine the additive treatment necessary. For low temperature operations, the pour point of the oil is perhaps of more importance than its oxidation stability.

The viscosity of the base oil in the grease will have been selected to suit the intended application for the grease. In general, higher viscosities are used when increased loads or duties are expected to be imposed on the grease. As with oils, it may be necessary to include extreme pressure additives and also corrosion inhibitors. Solid lubricants, such as graphite and molybdenum disulphide, are sometimes included for certain applications; for instance, when bearings under heavy loads are subjected to frequent start and stop conditions.

Premium grade greases used in industry must have their qualities approved by the various bearing manufacturers. The three most important properties specified are mechanical stability, oxidation stability and wear prevention in bearings.

The mechanical stability and also the pumpability of a grease is influenced by its yield value. This is the minimum shearing stress which must be applied to the grease before any flow will occur. This means that at low rates of shear a grease behaves more like a solid with little tendency to flow. At higher rates of shear, the grease has a greater tendency to flow and behaves somewhat more like a liquid. Unlike most normal lubricating oils, greases do not obey Newton's law of viscous flow. They do not possess a constant coefficient of viscosity at a given temperature. The viscosity will vary depending upon the shearing force employed. Greases have only an apparent viscosity at a given temperature, which may be defined as the ratio of the shearing stress to the rate of shear, at a given rate of shear value.

The selection of the correct consistency, or softness, of a grease for a particular application will be of great importance with regard to efficient lubrication. The consistency of greases is normally classified by a grade system suggested by the National Lubricating Grease Institute of the United States of America, commonly referred to as the NLGI. (The worked penetration experimental test method used to determine grease consistency will be described in the section on Physical and Chemical Laboratory Testing in a later chapter). The normal arbitrary numbering scale devised for most common greases will range from 0 (for soft greases) to 5 (for stiff greases). However, beyond the extreme ends of this scale range, there are also special additional grade numbers, 00 and 000 to denote semi-fluid greases and 6 to denote very stiff greases (Fig. 5.2).

Grade number	Structure	Penetration (worked) mm/10	Cone sinking depth
000	Semi-fluid	445-475	
00	Semi-fluid	400-430	
0	Very soft	355-385	
1	Soft	310-340	
2	Medium soft	265-295	
3	Medium	220-250	
4	Stiff	175-205	
5	Very stiff	130-160	
6	Very stiff	85-115	

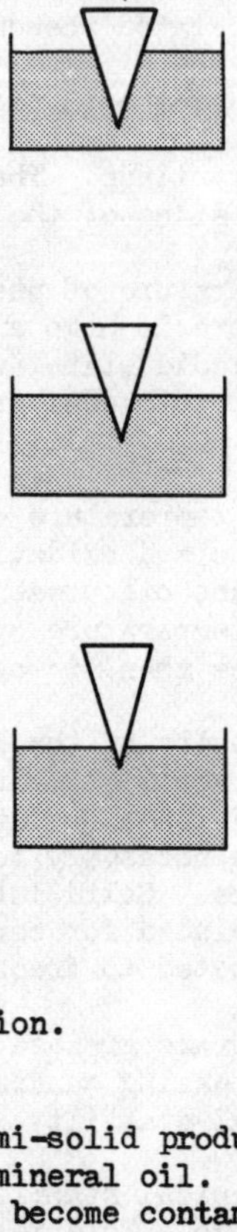

Fig. 5.2. Grease consistency classification.

The fact that grease has a consistency and is therefore a semi-solid product, means it cannot be handled in the same way as a liquid industrial mineral oil. In particular, great care has to be taken to ensure a grease does not become contaminated with dirt, sand or other grit. If this happens, the grease becomes abrasive to bearings and causes considerable damage. When the lid is removed from a grease keg, it must never be laid on the floor, especially with the face down, due to the grave risk of dirt being picked up on the tacky surface. For this reason, factories which only have limited off-takes of grease, often prefer to obtain their supplies in small packages or cartridges, from which the grease can be directly dispensed to the lubrication point.

Medium sized users of grease, normally employ special grease dispensing equipment to fill their own grease guns direct from the container. The dispensing equipment commonly consists of a grease pump. This is designed not only to avoid the pick-up of dirt during the filling operation but also to ensure that no air is allowed to enter the grease as it is being pumped into the grease gun. Other designs obviate the necessity to fill a grease gun, because the dispensing appliance is designed to fit directly on the top of the grease keg. The grease is then directly delivered to the lubrication point by a special flexible tube from the dispensing appliance.

A grease gun is therefore not needed.

Large users of grease, such as steel works, often supply the grease to a centralised lubrication system from a large storage grease hopper. The grease can then usually be delivered to the works in bulk. The hopper can normally hold several tons of grease and during use contamination is avoided by the incorporation of a plate which floats on the grease surface. The floating plate falls as the grease is used from the hopper. It is pushed up again, when the hopper is eventually refilled with a new supply of grease from the bottom exit of the hopper.

In practice, when grease is applied from any type of dispensing equipment to a bearing fitted with a grease nipple or cup, only a small amount must be injected at a time, otherwise severe overheating may occur. This is caused by frictional heat arising from the churning of the grease because the bearing is over-packed. To avoid a similar type problem, bearings which are pre-packed on assembly are usually only filled to about two thirds of their capacity with grease.

Outside their normal role of bearing lubrication, greases are also sometimes employed as a class of temporary corrosion preventives. The various different classes of temporary corrosion preventives form the next section of the chapter.

TEMPORARY CORROSION PREVENTIVES

The role of temporary corrosion preventives in the industrial oil field is to give short-term protection to metallic components or equipment. This protection may be during storage, transportation or between manufacturing processes. The name temporary implies the products are easily removable, when required, from the metallic surfaces. This is usually done by solvent or alkali degreasing. The products are therefore not designed for the same duties as the permanent protectives, such as paints and metal coatings, which are not intended to be removable after application. Before discussing the temporary corrosion preventives in more detail, it is worthwhile to examine the corrosion mechanisms they are designed to combat.

It is now established that corrosion of metals costs industry vast sums of money each year. The chief destructive mechanism is the atmospheric rusting of iron. Rusting is an electrochemical process and proceeds in the presence of air - providing oxygen - and water. Small differences in electrochemical potential are usually present on iron surfaces and these set up local anodes and cathodes (Fig. 5.3). In the presence of air and water, which acts as an electrolyte, the cathodic reaction which takes place on the surface produces rust. Rust consists of oxides, together with hydroxides of iron and their hygroscopic nature allows moisture to be trapped, which encourages further rusting.

In a simplified form, rusting is initiated at the local anodes, by iron going into solution in the form of ferrous ions

$$Fe \text{ (anodic)} = Fe^{++} \text{ (ferrous)} + 2 \text{ electrons}$$

After a while, polarization effects may cause the reaction to cease at an individual local site but it will restart again at new anodic sites on a continuous basis. The electrons left in the metal from the dissolution of each iron atom, flow to the local cathodes on the metal surface. At the cathodes, the electrons may react with hydrogen ions in the electrolyte, especially if it is of an acidic nature. Hydrogen gas may then be evolved by the following reaction

$$2H^+ + 2 \text{ electrons} = H_2 \text{ (hydrogen gas)}$$

If the electrolyte is not acidic, then the migrating electrons may alternatively

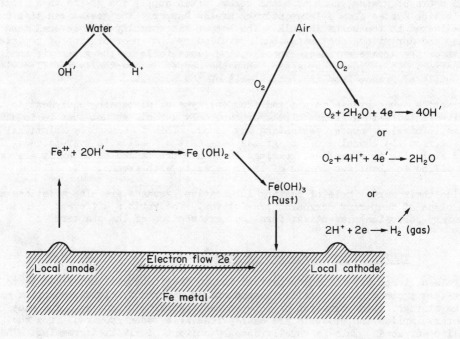

Fig. 5.3. Rust formation.

react with the oxygen from the air, which will also be present in solution in the water. The type of simple reaction which may occur with the formation of hydroxyl ions is then

$$O_2 + 2H_2O + 4 \text{ electrons} = 4OH' \text{ (hydroxyl)}$$

A further reaction can sometimes occur at the cathode which results in the formation of water by electron neutralisation of hydrogen ions

$$4H^+ + O_2 + 4e' = 2H_2O$$

The ferrous ions, formed in the anodic reactions, react with the hydroxyl ions, formed at the cathodes, with the formation and precipitation of ferrous hydroxide

$$Fe^{++} + 2OH' = Fe(OH)_2$$

The ferrous hydroxide precipitated on the metal surface may be further oxidised to ferric hydroxide, which is a simple form of rust. Rust may also be formed with a more complicated composition, such as hydrated ferric oxide or mixed oxides and hydroxides. The type of corrosion product formed will affect the kinetics of the corrosion process and modifications to the simple mechanism described may occur. Rust is hygroscopic in nature and attracts water, which further aggravates the rusting situation and leads to the establishment of an electrolyte film on the metal surface. The volumetric size of rust is greater than that of the iron from which it was formed and this leads to the creation of surface roughness. It also accounts for the formation of blisters on rusted surfaces which are painted.

The composition of a steel greatly affects the incidence of rusting and corrosion. Iron and steels have surface films of oxide formed on them as soon as they are exposed to the air. However, in the case of iron and most mild steels, this film is readily broken down and in the presence of water the film is not reformed and rusting occurs. In contrast, stainless steels possess a passive oxide film which is much more stable and less likely to break-down. This makes a stainless steel more corrosion resistant. Mild steels and iron surfaces are therefore the types most prone to corrosion.

The local anodes and cathodes present on iron surfaces can be initiated in several ways. The most common is by the presence of mill-scale (an oxide formed on the surface during the hot rolling of steel), which is cathodic with respect to the iron and therefore creates a possible condition for rusting to occur. Surface roughness, cracks and impurities can also induce local anodes and cathodes. The use of temporary corrosion preventives makes a barrier between the metal surface and the air and therefore inhibits the cathodic reaction which forms rust.

Mechanisms, other than rusting, can cause the corrosive destruction of unprotected iron and steel surfaces. The presence of sulphur dioxide and pollutants in the atmosphere can lead to the formation of acidic corrosion. Wood acids, exuding from wooden packing cases in contact with metal, can also cause a similar form of attack. Even mineral oils, used as a protective, can be oxidised when in thin films to form organic corrosive acids. Bacterial colonies present on the metallic surface can also set up a corrosive mechanism by the formation of oxygen concentration cells. The metal under the colony exists under anaerobic conditions and locally corrodes when it becomes anodic with respect to the colony edges. The edges have a higher concentration of oxygen and are cathodic. These forms of acidic corrosion are normally combated by the inclusion of basic inhibitors in the protective, to neutralise the acids as they form.

The storage environment plays a large role in reducing the severity of a corrosion attack. Outdoor conditions are obviously much more severe than indoor. Also, the corrosion of steel is much reduced if the relative humidity of the air is below 70 per cent and the presence of atmospheric pollution is minimal.

The temporary corrosion preventives are predominantly designed for the protection of ferrous materials under indoor, or outdoor short-term sheltered storage. They may be classified into three main types known as soft film, hard film and oil protectives (Fig. 5.4).

The soft film types frequently contain a solvent for ease of application of the protective film. When the solvent evaporates, the soft film is left evenly distributed on the metal surface. The film often consists of hydrocarbon material and natural products such as lanolin. Sometimes, these solvent deposited products possess so-called dewatering properties in addition, which means that metal components do not have to be dried before being dipped into the product. The dewatering grades have surface active agents incorporated in them, so that any water on the metal surface is displaced and the surface becomes preferentially wetted by the hydrocarbon material (Fig. 5.5). The displaced water falls to the bottom of the dipping bath where it is drained away at intervals (Fig. 5.6). In addition to the solvent deposited grades, there are also the non-solvent deposited soft film grades, such as the petrolatums and greases. Sometimes, the petrolatums have corrosion inhibitors incorporated in them to neutralise acidic corrosion. The petrolatums are normally heated before the components to be protected are dipped in them. The type of film formed is soft, thick and malleable. The exact thickness will depend upon the dipping temperature. This should not be excessive because if the petrolatum is overheated, then cracking

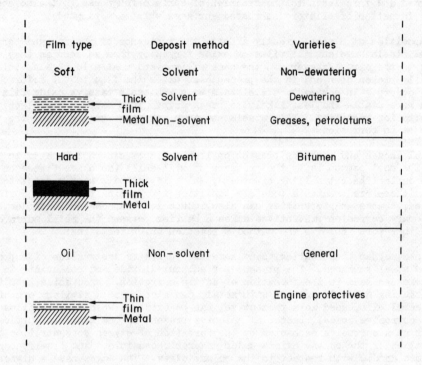

Fig. 5.4. Main classes temporary corrosion preventives.

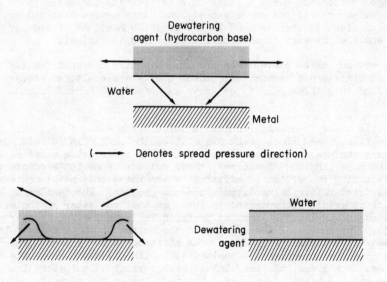

Fig. 5.5. Dewatering fluid action.

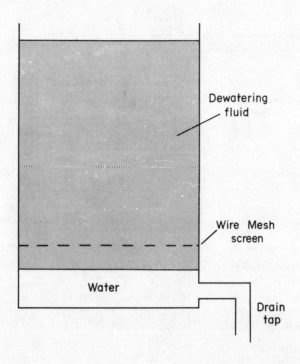

Fig. 5.6. Dipping bath.

is likely to occur with the possible formation of acidic oxidation products.

The immersion time in the bath will also affect the film thickness. It is possible to build up very thick films on the surfaces of the components to be protected, by repeating the dipping process. This increases the thickness of the film in stages, until the required level is reached. It is possible to apply petrolatums to components by cold smearing but care has to be taken to ensure that corners and edges are coated with an adequate film thickness (Fig. 5.7). It is preferable to apply petrolatums by the hot dip total immersion process. This ensures that complete and adequate coverage is achieved.

The thick film petrolatums can be used for the long term storage of components, under indoor conditions. They can also be utilised for storage under outdoor conditions, as long as a further protective wrapping layer is employed. This is ideally a grease proof paper. The additional wrapping protects the film from contamination and reduces the risk of mechanical damage. Roller and ball bearings are frequently protected during storage by the use of petrolatums. The thick film can easily be removed, when required, from the protected component. Besides wiping and solvent degreasing, a dip in a hot oil bath can also be used to remove the protective film.

The petrolatums employed as temporary corrosion preventives are normally green in colour. They are fibrous in nature and possess a grease-like structure. Petrolatums are also similar to conventional greases by the fact they possess definite drop points. Greases, particularly calcium based ones, are also frequently used as soft thick film protectives but are usually applied cold. The type of film formed is not as

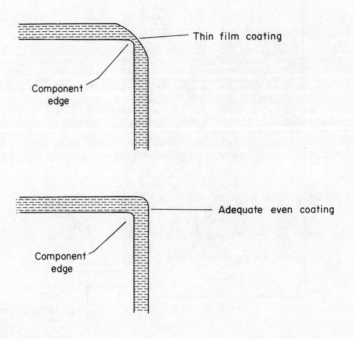

Fig. 5.7. Edge coating.

robust as the petrolatum type but has the advantage of reforming more readily, if broken by mechanical damage. In general terms, the solvent deposited soft film types are used for the protection of simple components whilst the non-solvent types are used for assemblies; especially if the assemblies contain confined spaces or components, such as rubber, which are attacked by solvents.

The hard film protectives are the second type of temporary corrosion preventives. They are solvent deposited grades which yield, as the name implies, a hard rather than a soft film, after application. These products are frequently based on hard film forming ingredients, such as bitumen, contained in a solvent. They protect metal surfaces for much longer periods than the soft film types, because the hard film is tougher and more resilient. They are used in such applications as car underbody sealants and for the protection of certain deck areas of ships.

The third type of temporary corrosion preventives are the oil protectives. These do not contain solvents and consist of mineral oil with corrosion inhibitors to combat acidic corrosion. They are used mainly for the protection of small components. Due to the relatively low viscosity of mineral oil, the films formed tend to be of a thin nature because of the oil drainage which occurs from a component after dipping. They give, therefore, less protection than the soft and hard film type protectives. A special class of temporary oil protective is used for the filling of gear boxes and crankcases of internal combustion engines. These oils are used for protection during transportation of the units, and are designed also for the units to be run for a short time on the oils, before filling with the service oil.

It will be appreciated that due to the different film thicknesses formed by the soft film, hard film and oil protectives, the products will be able to cover various surface areas. For example, the extent of the covering power of a solvent deposited product may be six times that obtained with the same volume of a hot dip product. This difference may be a factor in determining the most economic product to use, when an application allows such a choice to be made. The preparatory treatment given to the article to be protected is also significant in this respect. The cleaner the surface, the longer will be the protection period achieved by the application of the protective.

In service use, the solvent type protectives tend to lose solvent by evaporation and the dipping baths should be kept covered to minimise this loss. The evaporation causes the product to thicken and after a while it may become necessary to thin the product, thereby reducing the solids content, by the addition of further solvent. The use of solvent containing grades presents a fire hazard and so care must be taken in their use. Certain speciality oils are available, in which solvents have been replaced by water, to make fire-resistant products. These water extended protectives are used in the form of oil in water emulsions. The emulsions are frequently kept warm, so that after a component has been dipped in the emulsion, the water can evaporate more quickly from the film on the metal surface, leaving the protective material.

There are other speciality temporary protective oils which contain volatile, or vapour phase, corrosion inhibitors which are useful for the protection of the metal present in the air spaces of assemblies which cannot be wetted by the oil. There are also solvent deposited protectives which give rise to thin, transparent plastic coatings when applied to metal surfaces; these are especially useful for the protection of chromium plated parts. In addition, there are hot dip, transparent plastic coating types. These are frequently cellulose based and are used to protect larger assemblies during transit. The plastic cocoon formed around the assembly is removed, when required, by cutting and peeling.

In the field of steel rolling, special sheet coating oils are used for the protection of the rolled strip after tempering. These oils are used to protect the coiled strip during its transportation from the steel mill to the customer. These types of oils are usually formulated to suit the specified requirements of the customer. A motor manufacturer may require special degreasing properties for the oil, so that it can readily be removed by the established process used at the factory. Ease of removal of the coating oil is of prime importance in this case, so that the process of metal phosphating and the application of permanent protective paint coats can be readily carried out when desired. Before the transportation of the oiled coils of strip from the steelworks, it is normal practice to treat the exposed edges of the coils with additional protective. Edges are particularly prone to corrosion and are subject to rubbing during handling and transit. As an extra precaution during transportation, the coils of strip may also be protected with a wrapping of waxed paper.

Inside the tightly rolled coils of strip, air has been excluded by the mechanical action of the coiling process and therefore anaerobic conditions exist. Anaerobic bacterial corrosive attack can sometimes occur under these circumstances and the sheet coating oil may be called upon to neutralise such an attack. Bacteriacides are sometimes employed in the sheet coating oil for this specific purpose. It has already been mentioned that motor car manufacturers may stipulate a particular degreasing requirement for a sheet coating oil.

Users of other types of protectives, in other applications, may demand different individual requirements, in addition to the desired protective powers of the temporary corrosion preventive. Certain users may need a protective film which has a good flexibility under low temperature conditions. Some may stipulate a hard film type temporary protective which can pass a specified test for adhesion and stickiness.

Other users may desire a film which can stand up adequately under a constant high temperature storage environment. Sometimes, a drying time may be specified for a special application which involves a solvent deposited temporary corrosion preventive

The ability of a product to protect a metal surface from corrosion is normally demonstrated under laboratory conditions, using humidity or salt spray cabinets. However, storage conditions in practice are so variable that the results of laboratory tests can only give an indication of the expected performance in service.

It has been mentioned that the initial cleaning process carried out on the components to be protected is of paramount importance. The cleaner the surface, the longer will be the expected protection period for the temporary corrosion preventive. The surfaces to be protected may be contaminated with oil, grease, dirt, swarf, rust, scale, heat treatment salts or fluxes. Whatever the contaminant may be, it should be removed from the metal before the application of the temporary corrosion preventi

Oil, grease, dirt and swarf can be satisfactorily removed from a surface by the use of petroleum solvents, halogenated solvents or aqueous alkali detergent mixtures. Ultrasonic cleaning methods can be used to assist in the removal of fine solid dirt particles from the surface. The removal of rust can be carried out either by the use of mechanical or chemical methods. The mechanical treatment may entail shot or sand blasting the surface. The chemical treatment method may involve acid pickling, or an electrolytic alkaline rust removal process. The removal of heat treatment salts and fluxes from a surface may be carried out by water washing or acid treatmen

The type of treatment necessary for an individual surface will depend upon the contaminant present and also the surface finish, size, shape and other critical features of the component. For example, it would not be prudent to use the sand blasting technique on a finely finished surface. Acid treatment must be carried out with caution, if a danger exists of hydrogen embrittlement of the particular metal. Solvents would not be used in cases where the object, for example, possesses a component made of rubber or a painted surface is present.

CHAPTER 6

Cutting and Metal Working Fluids

CUTTING FLUIDS

Most of the oils discussed in detail so far, have been designed for the lubrication of industrial machinery. We will now consider some different types of oils used for other purposes. As mentioned in chapter 1, cutting fluids are classified as production oils. The fluid's main function is to assist machining operations, whereby a tool produces a component by a cutting action on a metal workpiece. The cutting fluid's role is twofold - it acts as a coolant and as a lubricant.

In a metal cutting operation, a tool shears the metal and the sheared metal removed from the workpiece forms into either continuous or discontinuous chips (Fig. 6.1). The energy resulting from the shearing of the metal is dissipated through the workpiece and tool, in the form of heat. Additional frictional heat is also produced by the flow and rubbing of the metal chips, as they are formed, over the surface of the cutting tool. The total heat released may cause the building up of some sheared metal on the tool surface, a phenomenon known as a built-up edge. This welding of tool to workpiece can be avoided by the rapid removal of the heat evolved and also by decreasing the total amount produced, by reducing the frictional heat component.

A fluid sprayed into the cutting zone achieves this aim. Sufficient heat can be removed, despite the fact that it is not possible for the fluid to penetrate into the precise region of the cutting edge of the tool, as it is completely enveloped in metal. Also, the fluid cannot, of course, penetrate into the body of the metal undergoing plastic flow ahead of the cutting tool tip. However, a copious well directed supply of cutting fluid can remove sufficient heat by metal surface cooling, as the fluid can penetrate fairly well into the region where the formed chip is rubbing over the tool, producing the frictional heat. The fluid can also lubricate the passage of the chip over the tool. The two main requirements for cutting fluids are, therefore, the ability to maintain the tool and workpiece at acceptable temperature levels and to reduce the frictional heat formed during the cutting operation.

The correct use of cutting fluids allows increased rates of production to be achieved in workshops. This is due to the increase of tool life obtained by reducing the tool wear, improving the stock removal rate, making power savings and obtaining better component surface finish, with more accurate dimensional tolerances. A further advantage of using a cutting fluid is that with ferrous components the residual fluid, remaining on the surfaces after the machining operation, prevents rusting occurring

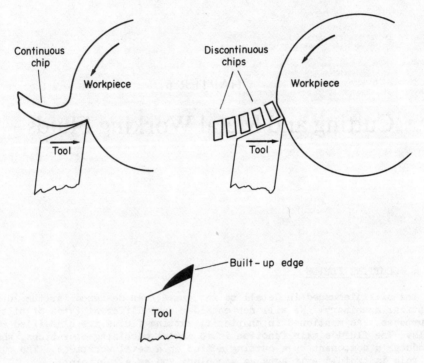

Fig. 6.1. Chip formation.

during short-term storage.

Three main types of cutting fluid are met in industry. The soluble oils are used as emulsions of oil in water and are the most widely used of the water based types. There are also water based synthetic fluids containing predominantly chemical additives, rather than oil. The additives are in true solution with the water and are not dispersed in emulsion form. The concentrate is diluted with water before use. The third main type is the neat cutting oil range, some of which include extreme pressure additives and fats (Fig. 6.2).

Emulsions of soluble oil, when prepared in water, are of a milky or clear appearance. This will depend upon the degree of dispersion, or the size of the oil particles, present in the continuous phase of the emulsion. In general terms, the greater the amount of emulsifying agent present in the soluble oil, the more clear and transparent will be the emulsion prepared from it. Emulsion stability is of great importance in service and the selection of the optimum emulsifier system for the particular oil used is a prime consideration. Also, the oil must be able to produce stable emulsions in the waters of various degrees of hardness met in industry.

The main function of a soluble oil emulsion is to cool the tool and workpiece; this is done efficiently, mainly because of the high specific heat of the water present. Soluble oil emulsions, because of their cooling power, are ideally suited for use in rapid and light machining operations, such as turning, drilling and grinding. However, it is possible to include extreme pressure additives in soluble oils to increase their range of application. The presence of the dispersed oil in the emulsion has some lubricating power but the primary characteristic of the soluble

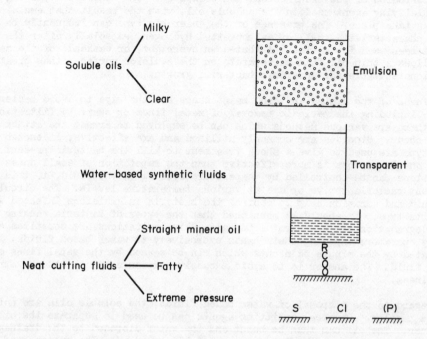

Fig. 6.2. Cutting fluids - main types.

oil emulsion is its cooling ability. The concentration of the soluble oil, dispersed in the water, will depend upon the individual application. It may range from 1 part of oil to 10 parts of water for turning, to 1 part of oil to 50 parts of water for grinding. For operations, such as grinding, it is important that the machinist can have a clear view of his work, as it progresses. It is, therefore, common practice to use transparent soluble oil emulsions for this application, or alternatively, the water based synthetic cutting fluids.

The water based synthetic fluids were originally designed for this purpose. They were essentially water, containing chemicals in solution to prevent rusting occurring during use. Later developments of the fluids, by the incorporation of lubricity and water soluble extreme pressure agents, have allowed their application to extend to more arduous duties. It is essential that both the soluble oil emulsions and the synthetic water based fluids do not cause corrosion or rusting of the machine tool or the workpiece.

Water based fluids are subject to bacteriological attack. The synthetic water based fluids are also prone to fungi attack. The presence of hydrocarbons, water and often nitrogen, sulphur and phosphorus compounds makes an excellent diet for bacterial growth. Initially, bacterial infection of the aqueous cutting fluid is usually caused by airborne dust, or the water used to prepare the emulsion. Once established, bacterial growth rates can be very rapid. Sometimes, the machine tool may not have been cleaned effectively, before the introduction of the cutting fluid. Stagnant pockets of a previously infected emulsion may be left behind. The bacterial attack may be of the aerobic type when air is present, or the anaerobic type in the absence of air. Aerobic bacteria frequently produce acidic components which can cause corrosion of the machine tool and workpiece. The anaerobic type can attack the emulsifying agent used in the soluble oil, with the result that emulsion breakdown can take place. The presence of the anaerobic type can frequently be detected by the characteristic smell of sulphuretted hydrogen, especially after the fluid has not been used for a time. A shut-down overnight, or weekend, of the machine shop allows a thin oil film to separate on the emulsion surface, thus creating conditions suitable for anaerobic bacterial growth.

Cleanliness of the system and good maintenance are the ways to avoid bacterial attack; including the periodic removal of metal fines or swarf by filtration. However, there are various methods which can be employed to arrest the attack if it should occur. Biocides are commonly utilised and are effective, if enough of the right type are used to give a shock treatment to kill the bacteria present. The use of one large dose is more effective than the repetition of small doses. Bacteri populations can be controlled by temperature cycling of the fluid, if this is possib Different bacteria thrive or die at various temperature levels. The circulation of the fluid and aeration help to control the bacteria in emulsions infected with the anaerobic type. It should be mentioned that the types of bacteria causing cutting fluid degradation are not harmful to man. Skin irritations are sometimes encountered by operators exposing their hands excessively to water based fluids. This is aggravated by the minute skin cuts which can be caused by the metal fines suspended in the fluid. The answer is to avoid excessive exposure and to observe strict cleanliness.

With regard to the disposal of water based fluids, the soluble oils are not too great a problem. Chemical splitting agents can be used to separate the oil from the water. The oil can then be burnt and the water disposed to the drainage system. This assumes that the water is of an acceptable standard. The presence of water soluble biocides may necessitate caution in certain cases. The synthetic water based fluids are not emulsions and therefore cannot be split. They are normally treated by diluting with large quantities of water before disposal.

The neat cutting oils are used for the slower and more difficult machining operatior such as gear cutting, screwing and broaching. The main ability required is lubrication to reduce frictional heat and thus decrease tool wear. Complicated tool form regrinding can be an expensive operation and therefore reduced tool wear can be a key factor in the economy of the machining operations. The neat cutting oils fall into three main classes, straight mineral oils, mineral oils blended with fatty oil and extreme pressure additive oils.

An especially important characteristic for the straight mineral oil class is the viscosity level chosen for a particular application. Although the oil must be able to lubricate effectively, the use of a low viscosity oil will improve the cooling ability which, of course, is advantageous. On the other hand, higher viscosity oil would have better retentive properties on the tool and workpiece in the region of the cutting zone. This is an important advantage in the slow speed cutting of the tougher metals.

Mineral oils, blended with fatty oils, are sometimes used when additional lubricati characteristics are desired. The fatty component has good friction reduction properties, due to the tenacious films it forms on metal surfaces. The compounded oils are also useful in the machining of metals, where staining by the cutting fluid may be a problem. Examples are the yellow metals (copper alloys) and aluminium alloys, which can be machined with compounded oils to give excellent surface finishes and minimal tool wear. The main disadvantage of compounded oils is that the fatty component is prone to oxidation; with the result that the viscosity and acidity of the oil may increase.

The extreme pressure neat cutting oils are based normally on sulphur or chlorine as the active components; although phosphorus compounds are also used to a limited extent. The reaction of these elements in the cutting zone, when tougher materials are being machined, prevents metal welding occurring under the boundary conditions which prevail. It is, of course, essential that the active agents are only released at the high temperatures existing in the cutting zone; otherwise, corrosion of the machine tool or workpiece would be likely to occur. The sulphur used may be in the form of elemental sulphur but it is more frequently in the form of a sulphurised fat, with the added advantage of the lubricity of the fat. The chlorine is normally in the form of a chlorinated hydrocarbon. The use of both sulphur and chlorine, combined in a fat, is also encountered and the two elements appear to have a synergistic effect. It will be recalled that sulphur is effective at a higher temperature level than chlorine. The use of a sulpho-chlorinated oil, therefore, provides for a wider range of cutting conditions.

Care has to be taken in the selection of the type of extreme pressure cutting oil for a particular application. For instance, elemental sulphur causes staining of yellow metals, such as copper based alloys, and therefore the use of a sulphur containing oil must be avoided. Heavily chlorinated cutting fluids must also be used with caution, because under certain conditions, hydrolytic decomposition of the chlorinated additive can form corrosive acids which may attack the ferrous parts of the machine tool. The use of this type of fluid is common for the broaching of tough high tensile materials.

In all cutting oil applications, whether with neat or water based fluids, it is important to maintain a copious supply of fluid, to the cutting zone. This is especially important when ceramic or cemented carbide tools are used. An interruption in fluid flow will allow large temperature variations in the tool, with the possibility of cracking of the tool tip and its early break-down.

The ceramics consist of various refractory oxides, such as aluminium oxide, which have been cemented together. In contrast, the cemented carbides are mainly tungsten carbide or titanium carbide, mixed with other alloy elements, set in a cobalt matrix by a sintering process. The ceramics are harder than the cemented carbides but are very brittle. However, the ceramics are cheaper and the tips are of the "throw away" type. Both the ceramics and cemented carbides possess low tensile strengths and special precautions have to be taken in tool preparation. The materials are normally used only in the form of tips which have been brazed or clamped to different tool bodies (Fig. 6.3). The cemented carbides are normally brazed, whilst the ceramics are clamped to the tool body so that they can be of the "throw away" type.

Unfortunately, the most desirable tool geometries for the most efficient cutting actions cannot be employed with the carbides and ceramics, because low strength tools would result. For practical purposes the tools have to be ground with the minimum possible clearances and top rake. This non-ideal situation results in the production of additional frictional heat during metal cutting operations. These processes are normally of a high speed nature with high production rates, as the properties of the ceramics and cemented carbides make them ideal for this sort of application. All these factors make it essential to use a water based coolant, which is efficiently and copiously applied to the cutting zone. Efficient heat removal prolongs the tool tip life and also protects the components being machined from undergoing thermal distortion.

In all high speed cutting and production rate processes, it is inevitable that large quantities of metallic swarf will be formed. This is a secondary reason for the correct application of the water based coolant to the cutting zone. A well directed coolant stream will efficiently flush away the swarf. The surface finish of the component being machined will be much improved by the flushing process.

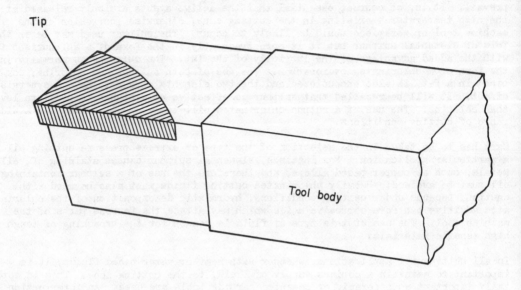

Fig. 6.3. Tipped cutting tool.

When swarf is removed from a machine, it inevitably takes with it some of the cutting fluid from the system. This "drag out" loss is relatively small with a water based coolant but it can be appreciable with a neat cutting oil, due to its greater viscosity. It is not normally an economically viable proposition to recover a water based fluid from swarf, for recycling back into the machine. In the case of a neat cutting fluid, the situation is different and it is sometimes economic to do a recovery process. The swarf coated with "drag out" oil is first allowed to settle. It is then passed through a course strainer to remove any remaining large metal particles. The separated oil is next frequently filtered through a magnetic separator, before being centrifuged to remove the remaining swarf fines. The oil is often heated to accelerate the various separation processes. The recovered purified oil is then returned to the machine, frequently as a "top-up" for the existing oil charge.

The policy of re-using neat cutting fluids, although often economic, is dubious on health hazard grounds. There is a possibility of the formation of minute quantities of carcinogenic hydrocarbons, such as the polynuclear aromatics, in a repeatedly used oil. It is feasible that the thermal cracking of cutting fluids, which can occur at tool surfaces, may produce such products. The longer an oil is used in service and the more severe the conditions, then the greater is the risk of excessive thermal cracking of the hydrocarbons occurring. There is therefore a risk that small amounts of carcinogenic compounds may accumulate in a neat cutting fluid, although the new unused fluid may originally have had potential carcinogens predominantly removed by a refining process. The activity of any carcinogen formed in a used cutting fluid, will depend not only upon its concentration but also the chemical structure of the precise aromatic compound. The presence of additives in the oil may also affect the situation.

Great attention has been paid to the safety of cutting fluids and their usage since it was realised that skin cancer was sometimes encountered on workers who allowed their skins to be exposed to such products in the metal working industry. However, the number of incidences has been rare when the total number of workers and the huge volumes of cutting fluids used in industry are taken into consideration.

Excessive skin exposure to many fluids may give rise to skin disorders. These may not be carcinogenic in nature. Dermatitis is perhaps one of the most common met in industry. Excessive exposure to oil may also give rise to the formation of oil acne, due to the blockage of the sweat glands. The formation of warts on the skin of workers in industry is also sometimes encountered under certain conditions. These are not usually malignant but occasionally they may become so.

It has been mentioned that water based fluids can sometimes cause skin disorders, if excessive exposure takes place. This applies more to neat cutting oils, especially as they are not soluble in water and therefore are less easy to wash off the skin. Also, the concentration of oil on the skin may be greater as they are used in an undiluted form. It is therefore very important that appropriate protective measures are taken to avoid skin contact and hygiene standards are kept high. The base oils used for preparing cutting oils are now refined to reduce the concentration of condensed hydrocarbon ring compounds, in order to minimise the presence of possible carcinogens.

METAL WORKING FLUIDS

Rolling Oils

It should be noted that outside of the field of metal cutting there are other metal working processes which entail no metal cutting action. The rolling of metals to reduce their thickness is perhaps the most important. In this case, metal is deformed but not cut (Fig. 6.4). In the rolling process, the same volume of metal leaves the rolls as enters it and therefore the speed of exit of the metal from the rolls is greater than the speed of entry. Some slippage therefore occurs as the metal passes between the rolls. The main properties desired of a rolling fluid are to control the amount of slippage, withstand the high roll pressures, cool the rolls and produce a good quality surface finish to the rolled strip. In this field of application, both emulsions and neat rolling fluids are used. In the steel industry, the type of fluid used will depend upon the thickness of the strip being rolled and the mill type. For the production of very thin strip, vegetable oils such as palm oil, have been traditionally used in admixture with water. Soluble oil emulsion type rolling fluids, with added extreme pressure and lubricity agents, are employed for cold rolling thicker strip sections.

In the aluminium industry, soluble oil emulsions are used for the hot rolling process. Ideally, the emulsions are neutral (pH 7), in order to avoid any chemical reaction with the aluminium. For the cold rolling of aluminium, fairly volatile narrow cut hydrocarbon fractions are used, together with polar additives. An important requirement is the avoidance of staining of the aluminium; the use of a volatile fraction ensures that the rolling fluid will volatilise from the metal surface during the annealing process, this reduces the possibility of staining occurring. Another very important factor with regard to the prevention of staining, is the efficient removal of fine particles of aluminium metal and other contaminating dirt from the cold rolling fluid. In practice, this is achieved by circulating the rolling fluid from the rolls into a so-called dirty tank, where the heavier contaminants can

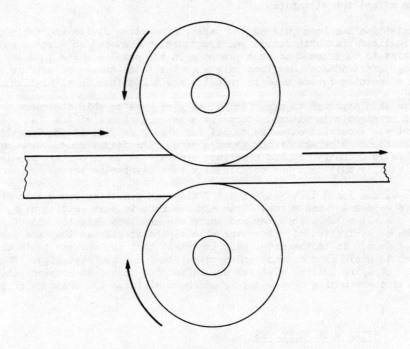

Fig. 6.4. Metal rolling.

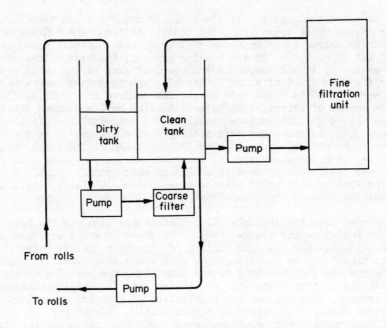

Fig. 6.5. Aluminium rolling oil filtration.

settle out (Fig. 6.5). The rolling fluid is then pumped into a clean storage tank, through a coarse filter to remove more of the remaining solid contaminants. The fluid in the clean tank is next pumped, either partly or fully, through a fine filtration unit to remove the remaining very fine contaminant particles. The fluid from the clean tank is then circulated back to the rolls of the mill.

Outside the field of metal rolling, there are other important industries concerned with metal working. These include wire drawing, cold extrusion and deep drawing.

Wire Drawing

Wire drawing bears some similarity to cold rolling, by the fact that the same volume of metal leaves the die as enters it and metal deformation takes place with some slippage in the die (Fig. 6.6). The speed of exit of the metal from the die is greater than the speed of entry, because the wire drawing operation reduces the cross-sectional area of the wire. The exit speed may be several hundred feet per minute, many times the entry speed into the initial die. In a wire drawing train, the wire is pulled through a series of dies so that the diameter of the wire is progressively reduced. Between each die the wire is passed around rollers to obtain the desired tension.

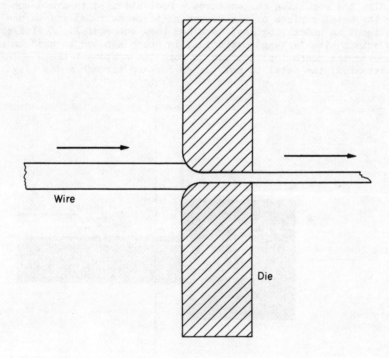

Fig. 6.6. Wire drawing.

The art of wire drawing is a complex phenomenon and much depends upon the skill of the operators. The lubrication of the wire during its passage through the die plays an important role, particularly with regard to lessening the amount of die wear. It is not always clear whether the mechanism involved is that of boundary or hydrodynamic lubrication, or a mixture of the two types. Much will depend upon the metal being drawn, the speed and reduction which can be achieved without causing surface cracking or significant changes in the mechanical properties of the wire. The pressures and temperatures in the die will normally be high. Many different lubrication products have been utilised but probably the most successful have been sodium based fatty acid soaps. The soap is applied to the ferrous wire before drawing, but usually after it has been cleaned and possibly also phosphated. The phosphating of the metal surface provides a good key to which the soap can adhere. Iron wire may be coated with lime before drawing dry with a soap. It can also be drawn wet, by passing the lime coated wire through an emulsion containing an appreciable quantity of a fatty material.

Cold Extrusion

The use of phosphate coatings and soaps has also become the most widely used lubricant system for the cold extrusion of ferrous metals. There are two alternative methods, forward or direct extrusion and backward or indirect extrusion. In the process of forward extrusion, the metal is pushed through a die, when it is required to form it into a desired component shape (Fig. 6.7). It thus differs from wire drawing, where metal is continuously being pulled through a die. However, as with wire drawing, the pressures involved in the cold extrusion process are extremely high and also the resulting temperatures. Most attempts to avoid the necessity to phosphate the metal surface of the component to be extruded and to use only a lubricant, without an underlying key, have not been successful. A similar lubrication situation exists with backward extrusion, in which a punch is used to cause metal flow back over the punch tool surface to form the component shape. In contrast to forward extrusion, the metal is not pushed forward through a die (Fig. 6.8).

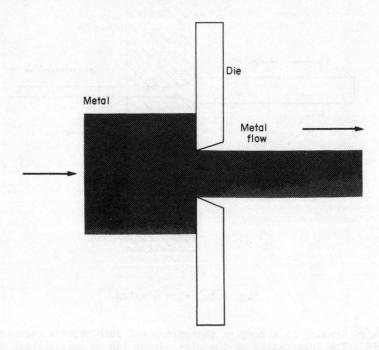

Fig. 6.7. Forward extrusion.

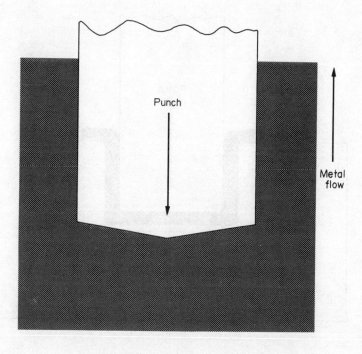

Fig. 6.8. Backward extrusion.

Deep Drawing

Conventional mineral oil based lubricants, such as sulphurised fatty products and other neat extreme pressure type cutting fluids, have proved successful in the process known as deep drawing. Graphite containing lubricants are also often satisfactory. It is not necessary to have a phosphate coating as a lubricant key. In the operation a metal blank, for example, may be pushed into a die by a punch to form a cylindrical cup (Fig. 6.9). In the process, metal is drawn over the die radius and into the wall of the formed cup. The severity of the operation will vary depending predominantly upon the amount of reduction achieved in a single punch operation and the composition of the metal being deep drawn.

ELECTRO-DISCHARGE MACHINING

Before leaving the various types of metal cutting and working processes, it may be of interest to mention briefly the process known as electro-discharge machining, or spark erosion. This method involves the controlled vapourisation of metal surface layers to obtain a desired component shape. A profile tool, which acts as the cathode is separated from the workpiece to be machined by a dielectric fluid, usually a low viscosity hydrocarbon mixture, such as a kerosine (Fig. 6.10). An electric potential is applied until the dielectric strength of the fluid is exceeded, so that a continuous spark or a controlled pulse electrical discharge occurs between the cathode and the workpiece, which acts as the anode. The temperature reached by the sparks are sufficient to cause vapourisation of a small area and depth of the workpiece on which they are directed. The metal is removed in such a way by the

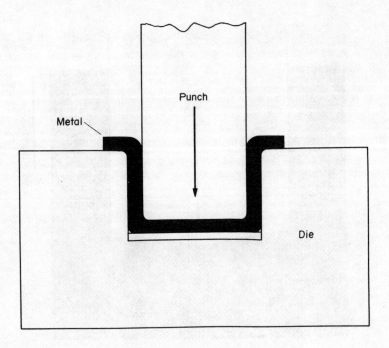

Fig. 6.9. Deep drawing.

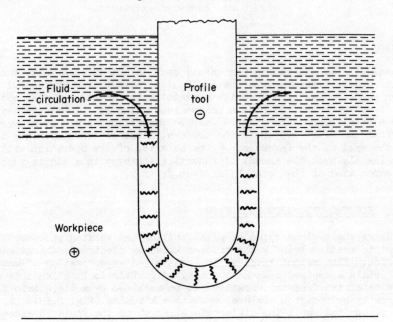

Fig. 6.10. Electro-discharge machining.

spark erosion that the workpiece eventually achieves the same shape as the profile of the cathode. The cathode tool is moved closer automatically to the workpiece, as the metal erosion process progresses. In this way, the distance across the spark gap is maintained.

The method is ideally beneficial for the machining of difficult tough metals, such as those frequently encountered in the aeronautical industry. The main function of the hydrocarbon fluid, other than acting as a dielectric, is to keep the gap between the anode and cathode clear of debris from the machining process. This is achieved by circulating the fluid through the gap to sweep away the debris as it is formed. The surface finish of the component is much improved by this process of continuous flushing and removal of metallic debris.

WORK HARDENING

In the general machining, cutting or cold working of metals, the ease of plastic deformation or the yield strength of the metal, will depend predominantly on the number of dislocations or imperfections existing in the metal crystallite structure. At the beginning of a metal working operation, the most easily moved dislocations will first shift. These slippages eventually become hindered in their movement by the dislocations building up behind impurities in the metal lattice. Other more difficult slip planes are then followed and disruption occurs. As the resistance to slip increases, the metal is said to suffer work hardening. At extreme levels of work hardening, no further slip in the metal may be able to take place and then metallic fractures are liable to occur.

Stainless and heat resistant alloy steels are especially prone to work hardening and are difficult to machine or metal work. Titanium alloys are similar in this respect and they are even more difficult to machine because they have a tendency to weld to any metal in sliding contact. In these cases of difficult to machine metals, it is normally necessary, as we have seen, to use neat cutting fluids containing high levels of extreme pressure agents. At the other end of the scale, there are mild steels which possess free-machining properties. Alloying elements are sometimes employed in the lower carbon content mild steels to improve their machining characteristics. Soluble oils are frequently used as coolants. Cast iron can be machined dry, as the free carbon present in the metal acts as an internal lubricant.

Outside the ferrous field, the main categories of metals which undergo machining operations are the copper based alloys, the aluminium alloys and the magnesium alloys. All can, in theory, be worked dry because they are easy to machine but in practice coolants are frequently used. In the case of magnesium, the main object of the coolant is to prevent the possible combustion of nascent magnesium surfaces in contact with the air. Low viscosity oils are frequently employed as the coolant. In the case of copper alloys, if a coolant is used, the avoidance of staining is usually a priority. Compounded oils containing a fatty component, are often utilised as previously mentioned. Soluble oils are frequently used for the machining of aluminium alloys, although they are also sometimes machined dry.

To generalise, the ease of machining of the particular metal plays a large role in the type of metal working lubricant required. This requirement ranges from nil for dry machining, through the soluble oil coolants to the heavily loaded, extreme pressure additive treated, neat metal working fluids. The severity of cut, or deformation, required by the machining operation also plays a role in the selection of the metal working fluid, together with the type of tool material used.

CHAPTER 7

Heat Transfer Applications for Industrial Oils

INTRODUCTION

There are many applications in industry where oils are used, not only for their lubrication properties, but also for their ability to transfer heat. In certain instances, the transference of heat may be required for cooling purposes but sometimes it is used for heating processes.

Heat Transfer in Bearings

In the common usage of industrial lubricants, it is worth noting that high temperatures can sometimes arise unnecessarily in a lubricant or grease. For example, this can be caused by the use of the incorrect grade, or quantity of lubricant, reaching the pressure contact zone of a bearing. The result is that effective lubrication and cooling by efficient heat transfer, with dissipation of heat from the bearing, cannot be achieved. During operation, a bearing produces heat due to a combination of the speed and the compressive loading. At high speeds, the temperature effect caused by the relative high motion of the contact surfaces, outweighs the heat contribution from the compressive loading. At low speeds, the compressive loading make the largest contribution to the temperature rise in the bearing. In bearing applications, the selection of the correct viscosity grade, as recommended by the bearing manufacturer, is of paramount importance. The first item to check, if a bearing has a tendency to overheat, is that the correct lubricant grade is being used and the correct quantity is reaching the bearing.

The viscosity of the lubricant at the working bearing temperature must, of course, be adequate for the build-up and maintenance of a hydrodynamic lubricant film. Also the quantity of lubricant, reaching the contact zone, must be sufficient to abstract heat quickly enough to prevent the viscosity decreasing below the required minimal value. However, the viscosity must not be excessive, as the additional energy required to overcome the internal frictional resistance of the oil, will result in extra heat being produced and retained in the lubricant and bearing. This additional energy is wasted energy and the power consumption will unnecessarily be increased.

The special importance of this aspect, in the textile industry, was mentioned in chapter 4, because of the vast numbers of spindles employed. It will be recalled that the viscosities of the oils selected are kept to the absolute minimum, to save power losses. In a few specialist applications in the textile industry, the use of air bearings has achieved some success. Gas under pressure, rather than a low viscosity lubricant, separates the moving surfaces in certain high speed applications. The viscosity of the gas is appreciably lower than that of a liquid lubricant.

Heat transfer effects are also important in the lubrication of rolling contact bearings. Ball and roller bearings are usually lubricated by grease, although in certain high speed applications, bearing oils are sometimes introduced to the bearings by one of the standard oil application methods. As with plain bearings, the higher the speed of the ball or roller bearing, the lower must be the viscosity of the oil to achieve adequate cooling and lubrication. In the more general and usual case of ball and roller bearings lubricated by grease, the quantity of grease packed or fed to the bearing is of extreme importance. As was mentioned in chapter 5, the normal recommendation is to fill the bearings to only two thirds of their capacity in order to avoid excessive heat production, due to churning effects and the creation of unwanted frictional heat in the bearing. The selection of the correct consistency and type of grease, also plays a major role in attaining optimum heat transfer and cooling characteristics, depending upon the designed operating temperature level of the bearing being lubricated.

It will now be a convenient point to discuss another aspect of heat transfer oils, as they are utilised in liquid phase heating systems. This application provides a good example of the thermal stabilities of industrial oils and their heat transfer properties under high temperature conditions.

Heat Transfer Oil Systems

Heat transfer oils are used in a wide variety of industrial applications. As the name implies, the oils are utilised for the transfer of heat from one place to another. In this application, the oils are not used for their friction reduction properties, or for assisting the production of components, but are used to convey heat for industrial heating purposes. Heat transfer oils are a good example of fluids which cannot be placed in one of the two rough categories, mentioned in chapter 1, of either a lubricant for industrial machinery or an industrial production oil. There will be other examples discussed in later chapters.

In a heat transfer system, a fluid is heated in a liquid phase heater, by a primary source of energy, such as electricity, gas or fuel oil. The fluid is heated as it passes continuously through the heater, on its way around an enclosed circuit. The hot fluid is pumped through pipes to the material to be heated; these pipes form a part of the enclosed circuit (Fig. 7.1). The heat is transferred, either by circulation of the hot oil through pipes in the material, or through the jacket of a vessel surrounding it. Besides the heater, pump, pipes and jacket, there is a necessity to include an expansion and a venting arrangement in a liquid phase heat transfer system. This is because fluids, when heated to the working bulk temperature, expand and increase their volume by up to 25 per cent. In an enclosed system, it is therefore essential that the fluid expansion can take place and this is normally done by the use of an expansion tank, connected to the main system, on the suction side of the pump, by a pipe of small diameter. This stops convection currents arising and allows the oil in the expansion tank to be kept cool. It is only in the expansion tank that the fluid is in contact with air. Vents must also be provided in liquid phase heat transfer systems, so that any inflammable vapours formed can be safely released to the atmosphere.

Before the advent of the liquid phase heater, steam was the main heating source used in industry and, of course, is still very widely used. Steam has many advantages but at higher temperatures, such as $300°C$, the steam pressures required in the system to maintain this temperature become uneconomic. It is in these temperature

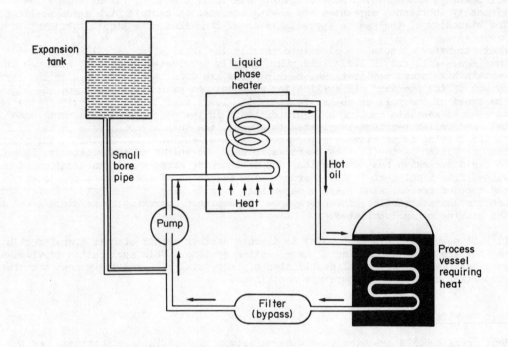

Fig. 7.1. Heat transfer system.

regions, in particular, that liquid phase systems have a great advantage as no pressures are required. The liquids are generally used below their boiling points and therefore require no pressurisation.

The type of liquid selected for liquid phase heat transfer will depend largely upon the bulk temperature required. Water possesses an excellent heat transfer coefficient, due to its high specific heat, but its temperature of use is limited by its relatively low boiling point of $100°C$. Above this temperature, of course, steam has to be used. Mineral oils find a much wider field of application, due to their high initial boiling points. They give satisfactory service in the temperature range of approximately minus $10°C$ to $320°C$. Still higher temperatures - up to $350°C$ - can be achieved with the use of highly aromatic hydrocarbons, such as the terphenyls and triaryl dimethanes. For temperatures up to $400°C$, eutectic mixtures of diphenyl/diphenyl oxide can be used. This chemical mixture can be utilised for heat transfer purposes in both the liquid and vapour phases. Its boiling point is $260°C$, so above this it is a vapour and is used like steam in this form. For temperatures in excess of $400°C$ and up to $500°C$, it is necessary to use eutectic salt mixtures, such as sodium nitrite and potassium nitrate. In the rare cases, when still higher temperatures are needed, liquid metals have an application to approximately $600°C$.

For extremely low temperature applications, it may be necessary to employ the organic silicate fluids, which have a very wide temperature range of fluidity from minus $50°C$ to $360°C$. However, these fluids are extremely expensive. The chlorinated hydrocarbons were at one time used to cover a wide temperature range from minus $30°C$ to $260°C$. Many of these fluids also possessed the added advantage of being fire-

Heat Transfer Applications for Industrial Oils 75

resistant. However, recent environmental objections against the use of chlorinated products has, as mentioned earlier prevented their present day acceptance. It is seen that many of the fluids overlap in a certain temperature range of application, which means a choice is available. In practice, industry would normally select the most overall economically available fluid to suit the particular application.

It is for this reason that mineral oils are the most widely used liquids in unpressurised heat transfer systems. They also have the advantage of not causing rust or corrosion in the systems. Most hydrocarbons start to thermally decompose at approximately $320°C$. It is therefore essential for the heat transfer system to be well designed to avoid overheating of the oils. The heater is a potential "cracker" of the oil, when the temperature of the fluid film at the heater tube surfaces is allowed to rise excessively. The film skin temperature in the heater is normally $30°C$ above the bulk temperature of the oil in circulation. In other words, if the bulk temperature is $320°C$ then the film temperature could be $350°C$. Heat transfer systems using oil are designed so that this film temperature figure is not exceeded. This is mainly achieved by ensuring that the fluid is in controlled circulation through the heater and is in a turbulent flow condition (Fig. 7.2). The use of high fluid flow velocities and the absence of any direct flame impingement on the heater tubes are essential. If the flow was allowed to become laminar or "streamline", then a stationary boundary layer of fluid would remain on the heater tube surface with the result that excessive "cracking" would take place. This would lead to carbon deposits which would impede heat transfer.

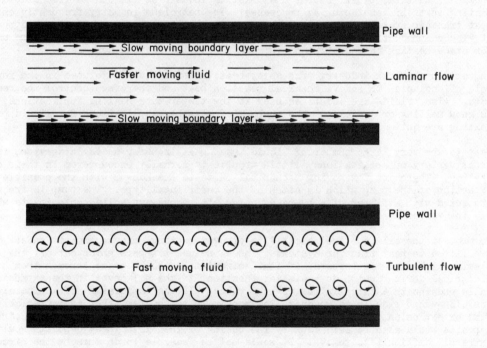

Fig. 7.2. Laminar and turbulent flow.

It will be seen that the two main requirements of a mineral oil heat transfer fluid are good thermal stability and a fairly low viscosity, so that it can be pumped readily through the heater, with turbulent flow characteristics. Oxidation stability should not usually be a problem, because in a modern designed enclosed plant the only contact between the oil and air is in the expansion tank. This tank should be kept at a temperature below $60°C$ to reduce the oxidation risk. The air in the expansion tank is sometimes replaced by an inert blanket of nitrogen, to give a slight overpressure; especially when the bulk fluid is being used at elevated temperatures of $320°C$.

Additives are not used in mineral oil heat transfer fluids because they would either decompose or volatilise at the working temperatures. It is advantageous for the oil to have a good viscosity index value, in order to facilitate ease of start-up of the plant before it reaches the working temperature. During the heating up process, the plant is constantly vented to remove any light hydrocarbon products formed. This procedure also removes air and water traces which may be in the system. The plants are also regularly vented, as a routine during operation, to ensure that any thermall cracked products formed are removed on a continuous basis.

The numbers and varieties of applications for heat transfer fluids in industry are immense. There are many applications where the indirect heating of a process by a hot fluid, rather than by direct flame heating, is advantageous. This is not only on the grounds of safety but also by the fact that a more accurate control of temperature can be achieved, as the risk of localized over-heating is avoided. Liquid phase heating systems are not only useful for manufacturing processes, requiring heat, but also for the handling and application of the product in its end use. The chemical, textile, paper, rubber and plastics industries all can utilise liquid phase heating in their manufacturing processes. The petroleum industry frequently employs heat transfer systems to assist in the handling, storage, transportation and pumping of viscous products, such as bitumen and heavy fuel oil. Hot oil can also be used for space heating in industry and for small district heating schemes.

In certain chemical industry processes, heat transfer fluids are used in the reverse role of coolants, to remove chemical reaction heat and therefore control the reaction rate. Also, fluids are sometimes used in low temperature heating applications, such as snow melting or de-icing. The numbers of possible applications for liquid phase heating are quite immense.

Despite the very large number of liquid phase heat transfer units in service, in a great variety of applications, little trouble is normally encountered in their operation. There are few moving mechanical parts to breakdown, with the possible exception of the pump which is often of the centrifugal type. The pump is frequently protected with a filter of a by-pass design, to remove any solid contaminants which may inadvertently have got into the system.

To protect the oil from the possibility of over-heating, many automatic safety device are fitted to heat transfer systems. These include triggers which cut off the electrical heaters or burners, if fluid circulation should cease through them. Also electrical devices which do a similar function, if the oil level in the expansion tan falls underneath a safe minimum pre-set value. If the temperature of a mineral oil, was allowed to rise in an uncontrolled manner in the heater, then an appreciable fire or explosion risk would exist. As mentioned earlier, large amounts of carbon deposits would also be laid down by the coking of the oil. These deposits would be extremely difficult to remove. It would not be feasible to do this by the circulation of solvents through the system and mechanical cleaning methods would have to be employed. When a heat transfer system is to be shut down, it is essential that the oil is allowed to circulate through the heater for a considerable time after the heater has been switched off. This procedure ensures the oil will not become overheated.

A rough guide to the interior state of a heat transfer system can be obtained by monitoring, at intervals, the property changes which may be occurring in the oil. Drastic changes in the open flash point, viscosity or neutralisation number are indicative of oil degradation and perhaps the need to change the oil. If any doubt exists as to the quality of the oil, then it is always best to change it. However, it must be first ascertained that the deterioration of the oil has not been caused by some malfunction or misuse of the heating system. A new oil charge will deteriorate by the same means, if any system fault is not first corrected. Mineral oils will survive for many years in well maintained and controlled heat transfer systems. It will be obvious that the lower the average operational temperature, the longer the oil will last.

When an oil change is to take place, it is essential that the heat transfer system is first thoroughly cleaned, before the introduction of the new oil charge. The use of a flushing oil will be beneficial for the removal of oil soluble sludges and varnishes from the system. Loose solid particles will also be removed in the flushing process. Finally, the system should be thoroughly tested for leaks before filling with the new oil charge. Water should never be used for the flushing process because it will not dissolve the hydrocarbon sludges. It would also be extremely difficult to drain the last remnants of water from the system. This would lead to operational difficulties due to the formation of steam pockets, when the system is filled with oil and heated to the operational temperature. Excessive venting would have to be carried out to dry the system.

In the special case, when a fire-resistant chlorinated hydrocarbon has been used as the heat transfer fluid in the past, it is advisable to take extra precautions. Chlorinated residues left in the system are potentially corrosive when they become hydrolysed and they will lead to the formation of strong acids which will cause corrosion. Such systems should be flushed with an oil, such as a detergent diesel engine oil which contains an alkaline additive, to neutralise the residual acids, before filling with a new mineral oil charge.

Electrical Oils

Electrical oils also transfer heat but they have a different function from the heat transfer oils just described. Electrical oils dissipate the heat formed in electric transformers, electric switches and motor starters. They also act as electrical insulators.

To ensure efficient heat dissipation, electrical oils always possess low viscosities, so that convective currents can readily be set up in the oils, in order to obtain the maximum cooling effect. Transformer and switch oils encounter very cold conditions, in units situated on exposed outdoor sites. The oil temperature becomes very low but it is essential that the oil retains good fluidity characteristics, so that its cooling ability is not impaired. For this reason, it is normal practice to use naphthenic type oils in this application, because of their low pour points and good low temperature properties.

The use of pour point depressants or other additives, such as oxidation inhibitors, is generally not favoured in electrical oils. This is probably due to the fear that poorer quality base oils could then possibly be used in service. Inferior base oils could lead to such detrimental effects as sludging, production of corrosive products, increase of oil conductivity, gas formation and the deterioration of the electrical insulator properties of the oil. Electrical oils are always well refined to ensure the long service lives required are achieved without the use of additives. Most countries and electrical authorities produce their own national specifications which rigidly lay down the precise type of oil required. A few of these do allow the presence of a selected oxidation inhibitor for certain applications, but the majority do not.

With regard to electric insulator ability, hydrocarbons possess high resistivity values. Few polar compounds remain in a well refined oil. Therefore, under normal conditions oils perform satisfactorily as dielectric fluids. However, under extreme high electric potentials, hydrocarbons can ionize momentarily and become conducting enough to allow a spark discharge to take place through them. After this, deionization occurs rapidly and the oil reverts to its former resistivity level; thus preventing a further discharge unless the potential is raised to a high value again. The high amount of energy released in a spark discharge can result in some permanent break-down of a hydrocarbon. A little gas formation may take place, together with varnish or sludge deposition. Therefore, the greater the stability of the hydrocarbons present in an electrical oil, the less is the amount of permanent break-down Good gas absorption characteristics are also favoured, to prevent the build up of inflammable gases in an electrical unit and the resulting potential fire risk.

Electrical oils must always be kept clean and very dry, in order to avoid any contamination causing a deterioration in the electrical resistance. Minute traces of contaminants in an oil can significantly increase its conductivity level. For this reason, electrical oils are always transported and delivered in vehicles which are segregated for this purpose. This makes it virtually impossible for the electrical oil to become contaminated with another product. Water contamination, due to its high conductivity, is a big problem with electrical oils and special vacuum treatments are used to remove the final traces. In a typical treatment, the oil is heated and sprayed into a unit, held under near vacuum conditions which results in the boiling point of any water present being reduced to $7^{\circ}C - 10^{\circ}C$. Under these conditions, the water evaporates immediately from the fine hot oil spray. The oil then possesses a very high dielectric strength, suitable for its intended duty as an insulant and coolant.

Due to the great care taken in the refining, finishing treatment and delivery of an electrical oil to avoid contamination, it is obvious that equal care must be taken in filling a system with the oil. Precautions are therefore always taken to ensure that contamination does not occur during the filling operation, or during its after service use.

The ingress of contaminants into a transformer will considerably shorten the service life, due to the deterioration of the dielectric. Power transformers are designed to minimise the possibility of contamination occurring, both during manufacture and service. If water should contaminate a transformer, then the traces of moisture can slowly diffuse through the transformer oil. The diffusion coefficient will be affected by the physical conditions prevailing, such as temperature, rate of oil circulation and the power load. The chemical state of the oil will also have an effect, depending upon whether the oil is new, or has become slightly degraded and oxidised during service.

A common method to assess the approximate condition of an electrical oil, with regar to contaminants, is to determine experimentally its electrical break-down value. The presence of a few parts per million of contaminants will result in a dramatic change in the break-down value. Before the test, it will be obvious that great care will have to be taken in obtaining and preparing the sample for test, due to its pronounced sensitivity towards the presence of contaminants.

Whilst on the subject of electrical oil, it will be appropriate to mention also the use of oils to fill electric power cables, rather than transformers or switch gear. In this cable application the oil is used primarily as an electrical insulator and thus a high resistivity value is again required. This will result in the oil possessing a low power factor and therefore low dielectric current losses. It is important that a cable oil does not precipitate wax during its service use, as this will increase the power factor. From the point of view of electrical properties, normal transformer oils are suitable as cable oils. However, a further problem which has to be contended with, is the formation of gases in the cable from the electric stressing of the hydrocarbons in the oil. An oil which possesses improved gas absorption characteristics constitutes an essential additional requirement of a cable oil. Other requirements for the oil are good oxidation stability, freedom from moisture and compatibility with the materials of construction of the cable.

Heat Treatment Oils

Oils used for the heat treatment of steels can be viewed as a further extension of the concept of a heat transfer application for an industrial oil. Steels are hardened by heating to high temperatures and then quenching them in a liquid to cool them rapidly enough to form martensite (Fig. 7.3).

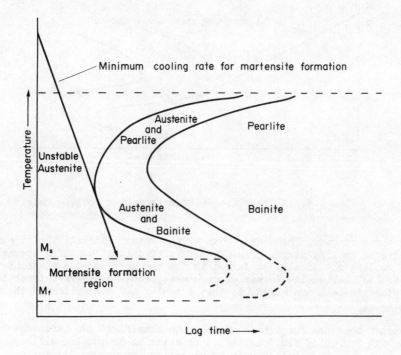

Fig. 7.3. Steel-time - temperature - transformation curve.

In general, steels can be classified as either water or oil hardening types. The precise alloy content largely determines the exact category, because it affects the quenching speed required for effective hardening (Fig. 7.4). With the oil hardening steels, the formation of thermal cracks is much less of a problem because slower cooling rates are involved. The larger alloy contents of these steels allows them to be satisfactorily hardened at the slower speeds, obtained by quenching in a mineral oil. The type of oil used is normally of a spindle oil viscosity. Too high a viscosity produces large drag-out losses when the steel is removed from the bath. The oil must possess excellent oxidation and thermal stabilities due to the severe conditions of use.

When a hot steel is quenched into a large quantity of cool liquid, the cooling takes

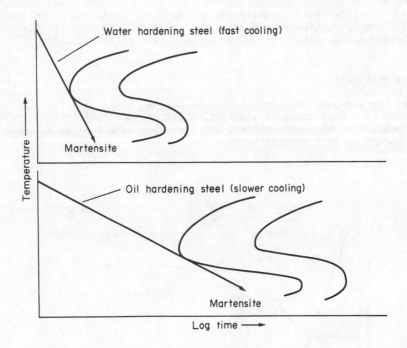

Fig. 7.4. Alloy effect - transformation cooling rate.

place in three stages. These are known as the vapour blanket, the vapour transport and the liquid cooling stages (Fig. 7.5). In practical quenching operations, fast rates of cooling are ideally preferred in the vapour blanket and vapour transport stages, so that martensite formation can take place. A slower rate at the liquid cooling stage is then preferred, so that cracking and distortion of the steel can be minimised (Fig. 7.6).

Factors other than the properties of the quenching fluid can influence the cooling rate. A high degree of fluid agitation by stirring or pumping will appreciably increase it. This is mainly due to accelerating the vapour blanket collapse. The fluid may be continuously cooled to lower its temperature, especially if high loads of steel are quenched in the bath. This cooling of the fluid allows, not only a faster rate of quenching to be achieved, but also a more controlled reproducible quench.

For the vast majority of oil hardening steels, straight mineral oils are entirely satisfactory to produce the desired hardenability characteristics. However, in certain cases (for example, when oil hardening steels are used containing the lowest concentration of their specified alloying elements) it is sometimes necessary to have a cooling rate slightly in excess of that obtainable with a straight mineral oil but less than that with water. The problem can be tackled in two ways. Additive can be used to increase the quenching speed of the straight mineral oil. Alternatively, materials such as alcohols, glycols and derivatives can be used to slow down the quench speed of water, so that the thermal crack problem can be avoided. The use of both types of quenchants is generally restricted to heat treatment of small section components. With heavier components, the heat retained in the bulk of the

Heat Transfer Applications for Industrial Oils 81

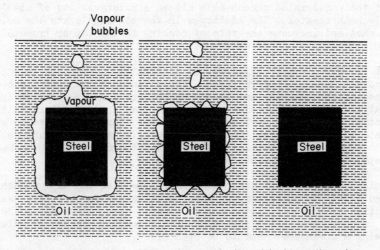

Fig. 7.5. Cooling stages.

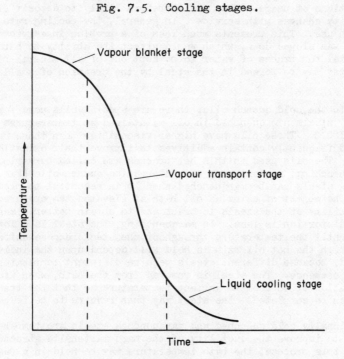

Fig. 7.6. Cold quench cooling curve.

steel tends to back-temper the surface hardness, thus losing some of the advantage gained by the fast quench. The thermal conductivity of the steel therefore limits the effect of a rapid surface cooling in large components.

The use of the accelerated quench oils allows a wider variety of steels to be satisfactorily heat treated. The additives in the oil accelerate the collapse of the vapour blanket and increase the rate of cooling in the vapour transport stage. The final liquid cooling stage is unaffected.

The use of the glycol-water mixtures, as alternative quenchants to accelerated oils, has both advantages and disadvantages. The advantages are that cleaner components are obtained after quenching and the quenchant produces no oil mist in the atmosphere of the heat treatment shop. Also, the product has a greater versatility, in that the concentration of the glycol and water mixture can be adjusted to give a selected cooling rate to suit the steel being quenched. The main disadvantage is that water evaporation occurs during service and the fluid therefore has to be monitored so that appropriate adjustments can be made.

Other speciality mineral oil quenchants exist, besides the cold quench and the accelerated oils. Emulsifiable, or water soluble quench oils, are examples and have surface active additives included in the mineral oil. These allow the oil residues, remaining on components after quenching, to be readily removed by water washing; thus allowing a clean component to be obtained.

Another speciality oil is required for vacuum quenching; where the steel is heated in a vacuum, rather than air, before quenching in the oil bath. Oils possessing defined vapour pressure characteristics are needed for this application.

Quenching oils of all types undergo thermal cracking and oxidation, due to the severe conditions of usage. They are therefore likely to deposit sludge and their cooling ability changes with service. In general, the cooling rate of an oil will increase with use. This presents much less of a problem in service than if the cooling rate was slowed down, which would affect the ability to harden the steel. It is essential for traces of water to be kept out of all mineral based quench oils, as "soft spots" may be formed in the steel by the creation of small localised steam pockets.

In addition to the cold quench oils, there are special oils used for hot quenching. In this case, steels are quenched in oil maintained at temperatures in the region of $165^\circ C$ to $180^\circ C$. These oils have higher viscosities than those used for cold quenching and frequently contain additives to improve their oxidation and thermal stabilities. The oils used in this hot process are called by various names, including marquenching, martempering and interrupted quench oils (Fig. 7.7). Most oil hardening steels can be marquenched when it is essential to produce an accurate component. The object of using an oil bath at elevated temperatures is to reduce the susceptibility of the steels to crack and to obtain better dimensional tolerance because the distortion is less. In marquenching, the steel is allowed to remain in the hot oil until the temperature throughout the steel becomes uniform. The temperature at which the hot oil bath is held will depend upon the individual steel being treated, because different steels require different temperatures for martensite formation to commence. The steel is removed from the bath, when its temperature is uniform, to allow it to cool to ambient temperature so that the transformation to martensite can be completed. The steel may then require to be tempered.

Both conventionally cold quenched and marquenched steels are tempered to relieve stresses and to improve the ductility of the hard martensite structure. When oils are used for this process, the bath temperature may be held in a region up to $280^\circ C$. This temperature level is far in excess of that used for marquenching. Tempering oils must possess higher viscosities, to obtain higher flash points and lower

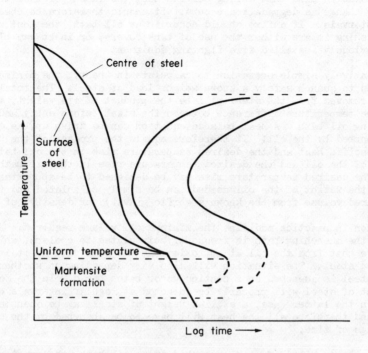

Fig. 7.7. Hot quenching - martempering.

volatility characteristics. Tempering oils are subjected to harsher conditions, which encourage more oxidation and sludge formation, than other heat treatment oils. Their service lives are therefore less than the other types of quenching oils.

The size of the component being tempered will normally determine the length of time it will have to be kept in the tempering bath, to achieve temperature equalization throughout its thickness. Mechanical stirrers, or other agitators, are not normally needed in a tempering oil bath, because the convective currents set up by the hot oil give sufficient circulation. After the required soaking time has elapsed, the tempered component is normally allowed to cool slowly to room temperature in the air. Steels which require tempering at temperatures higher than those which can be safely obtained with mineral oil, are normally tempered in fused salts or liquid metal baths. Alternatively, they may sometimes be tempered in electrically heated hot air circulation furnaces.

It will be obvious that in both tempering and quenching operations, there exists a greater fire risk than with many other industrial activities. It is thus very important, that a sufficient quantity of oil is present to cope with the proposed loading of hot steel into the bath, during a quenching operation. Otherwise, the oil temperature may rise above the fire point and ignition may occur. There are always potential ignition sources present in a heat treatment shop; perhaps the most likely ones are furnaces, or hot steel only partly immersed in the quenching oil. Closely associated with the total oil quantity is its rate of circulation, which can be adjusted to prevent local pockets of over-heating occurring in the oil bath.

In a quenching or tempering oil bath, the oil temperature must never be allowed to rise to a value exceeding $30°C$ below the open flash point of the oil. Also, it should be borne in mind that during service use, the flash point of the oil will appreciably change as degradation occurs. It should therefore be checked at fairly frequent intervals. If a fire should occur in an oil bath, the most usual methods of extinguishing it are either the use of tank covers, or inert gas blankets obtained from previously installed fire fighting equipment.

It is a relatively simple operation to calculate in theory, the minimum volume of oil required to quench safely a known weight load of steel. The total amount of heat to be removed from the steel will be the product of its weight, its specific heat and the temperature difference between the steel before entry and after leaving the quenching oil bath. A heat balance equation can be drawn up, as this total heat must be absorbed by the oil. It therefore equals the product of the weight of the oil, its specific heat and the desired temperature rise in the oil bath. The specif heat value of the oil and the desired temperature rise in the oil bath are known factors. The desired temperature rise may be designed to be approximately $20^\circ C$. Therefore, the weight of the oil needed can be simply calculated and then converted to the desired volume from the known specific gravity, or density, of the oil.

The situation in practice modifies the minimum oil volume required. For instance, the oil in the quenching bath is frequently circulated to coolers, which are designed to remove heat from the oil at approximately the same rate as it is gained from the quenched steel. The situation will also vary depending on whether a given weight of steel is quenched on a continuous or batch basis. In the former case, a given weight of steel will probably be quenched at regular intervals over a period of time. In the latter case, a similar weight of steel may be quenched in a single operation and therefore all the heat will have to be absorbed by the oil in a much reduced space of time.

The greater the quantity of steel quenched in an oil bath, the greater are the drag-out losses (the oil removed and lost from the bath by adhering to the quenched components). The viscosity of the oil also affects the drag-out losses. Higher viscosity oil baths will yield higher losses than lower viscosity oil baths. In any event, drag-out losses are normally high so the top-up rate with new oil is frequent, which assists in keeping the oil charge in good condition. However, it is inevitable that scale from the surface of the steel and sludge from the oil will gradually accumulate in the quenching bath. These can be removed at intervals, if they separate to the bottom of the bath. Filtration is normally also carried out to remove suspended contaminants. Caution must be exercised if activated earth filters are used, because they may also adsorb surface active additives which may sometimes be present in the oil and also generally affect the cooling rate of the quench.

As mentioned earlier, traces of water in a mineral oil quenchant must be avoided because they can cause "soft spots" on the steel. Larger amounts of water may even cause foaming to occur in the oil bath. In practice, the most usual source of water contamination is a broken cooler. It is essential that any water in the bath is successfully removed by separation. If this cannot be achieved, then it may be necessary to replace the whole oil charge in the system. The type of cooler used will depend upon the complexity of the quenching system. In simple designs, the cooler may be in series with the quench oil tank and the oil may be continuously pumped through it. In more complicated systems, where a storage supply tank may feed oil to the actual quenching bath, the cooler may be linked only to the supply tank and cool this on a recycling basis.

Under normal circumstances, the replacement of a quenching oil charge is fairly infrequent. With topping up, the same oil is often used for over a year. Oils employed for speciality vacuum quenching last considerably longer. This is because the steel is not only preliminarily heated in a vacuum but the actual quenching operation is also carried out under vacuum conditions. Oxidation of the oil does not therefore occur and the formation of sludge is greatly reduced. In contrast, tempering oils have to be changed fairly frequently due to the severe conditions which considerably increase their viscosities during use.

CHAPTER 8

Industrial Oils in Hostile Environments

In industry, oils are frequently called upon to operate satisfactorily under many different types of hostile environments. The conditions may range from extreme heat to extreme cold; the lubricants may be exposed to radioactive radiation in certain nuclear engineering applications, or high vacuum conditions in other industries. Sometimes, chemically corrosive atmospheres and also dusty abrasive conditions have to be tolerated.

High Temperature Environments

In many applications, lubricants must function under high temperature conditions, due to the very hot environments in which they are required to operate, such as those met, for example, in the steel industry. In other instances, the high temperatures may be caused by the high loadings and speeds imposed on the machinery in its service operation. A main requirement for an oil, to function satisfactorily at high temperature without lubrication failure, is that its viscosity does not decrease to such an extent that the oil film thickness falls below that required to prevent asperity contact in the lubrication zone. If this occurs, then wear will result. High temperatures, with the resultant thinning of the oil, may accelerate the change to a boundary lubrication condition. However, the correct initial selection of the oil viscosity, slightly higher than would be used under less arduous conditions, helps to overcome the problem.

The chief factor which really limits the use of mineral oil based products, at high temperatures, is the oxidation stability, rather than the viscosity thinning effect or the thermal stability. It is not advisable, in the presence of air, to use mineral oil based products in any application at temperatures approaching $200^{\circ}C$. With mineral oils, the higher the temperature the greater is the accelerated rate of oxidation; a small increase in temperature at a certain level (which varies with different oils) may result in a phenomenal increase in the oxidative break-down of the oil. Anti-oxidants present in the oil may delay the matter but they are not so effective at higher temperatures. In general terms, the lower the oil temperature, the longer will be the anticipated service life.

In most high temperature lubrication situations, it is advisable to avoid the use of mineral oils which are compounded with fatty materials. Fats are even more prone to oxidation than mineral oils; the deposition of sticky residues is therefore more likely to occur, due to the break-down of the fatty component.

High temperatures and the presence of air also enhance the potential fire risk with mineral oil based products. However, it has been seen that in the absence of air, when the thermal rather than the oxidation stability is the major factor, mineral oil based products can be stable and perfectly safe at temperatures in the region of $300^{\circ}C$, when used in specially designed enclosed heat transfer systems.

High temperature greases, based on mineral oil, must also not be used at temperatures much in excess of $200^{\circ}C$, due to the limiting presence of the hydrocarbon oil and its risk of oxidation. The types of grease formulated to be stable at these higher temperature levels cannot be thickened with conventional soaps, as the relatively low melting points of these would limit the attainment of a temperature approaching $200^{\circ}C$. Unfortunately, these non-soap greases are not suitable for the lubrication of bearings, except when they are running at relatively low speeds (below 1500 revolutions per minute).

However, faster bearings still have to be lubricated in hot environments. A good example of such grease lubricated bearings, operating under high temperature conditions and higher speeds, may be found in the hot roll necks of metal rolling mills. Some old rolling mills still possess bearings which have to be lubricated with blocks of grease placed on them and the grease slowly becomes emulsified with the cooling water. Even longer ago, hot bearings were lubricated and cooled by the positioning of a fat in a permeable bag on them, from which the fat melted and spread to the bearing. A stream of water helped the cooling process and prevented a too rapid melting of the fat.

Nowadays, high melting point grease blocks are available for use on rolling mill bearings with heated necks. A high temperature grease block operates satisfactorily at temperatures up to the region of $150^{\circ}C$. However, most modern precision rolling mill bearings are now mechanically lubricated with fairly high viscosity oils, as was described in chapter 3.

Fortunately, in the majority of cases, mineral oil based lubricants do not have to function at anything like their upper oxidation or thermal stability limits. Also, in many cases of high temperature operation, it may be possible to reduce the temperature of the oil by the installation of an air cooler, or other suitable heat exchanger.

In general applications, where temperatures in excess of $200^{\circ}C$ are anticipated, products based on synthetic lubricants, such as the diesters, or solid lubricants find their outlets and are extremely useful. In more specialised outlets, the organo-silicates are utilised as they are thermally stable to temperatures in the region of $350^{\circ}C$. However, prolonged usage at this high temperature level causes a certain amount of slow decomposition to occur and eventually the lubricant has to be renewed.

It will be obvious that water based synthetic products cannot be stable under high temperature conditions. It is never advisable to use such products much in excess of $60^{\circ}C$. If non-inflammability is a special requirement on safety grounds, due to possible lubricant leakage at high temperatures, then it will be necessary to employ a fully synthetic type of fire resistant lubricant, like a phosphate ester. However, it is not advisable to use such a fluid for any length of time above about $150^{\circ}C$. The silicates, of course, could be used at higher temperatures, if the price level was acceptable.

Low Temperature Environments

In the case of low temperature applications for industrial oils, the oxidative and thermal stability of the lubricant is obviously never such a significant factor.

The main problem under low temperature conditions is to keep the lubricant in a liquid state. The viscosities of oils increase dramatically at low temperatures, so it will be essential to employ an oil which has a relatively low viscosity at normal room temperatures, if it is to retain fluidity at much lower temperatures.

A low viscosity is always an essential factor for the efficient start-up of machines. This problem will obviously be aggravated at low temperatures, due to the slower oil flow rates and the slower lubricant spreading rates over metallic surfaces. Another problem, accentuated at low temperatures, is the higher degree of churning which can occur with the lubricant, due to its higher viscosity. The churning process may prevent an adequate amount of oil from flowing between the contacting surfaces, with the possibility of a lubrication failure occurring.

At first sight, it may appear advisable to employ an oil with a high viscosity index to reduce the change of viscosity with lowered temperature. However, superimposed on the viscosity effect is the additional factor of wax separation from the oil at low temperatures. Wax precipitation has the result of thickening or solidifying the hydrocarbon oil, so that it can no longer be considered to be in a liquid state. Low wax content oils are therefore beneficial for low temperature lubrication. It will be recalled that, due to this aspect, naphthenic oils are the usual recommendation for refrigerator oils, although their viscosity indices are usually poorer than most paraffinic type oils.

However, low temperature problems are not confined to refrigerator oils. Equipment of many types has to operate under cold outdoor conditions. Transformer and switch oils are typical examples and again naphthenic oils are the normal recommendation. The use of pour point depressants in other types of oil, such as the paraffinics, are essential if they are to achieve fluidity at low temperatures, so that lubrication failures can be avoided. Hydraulic systems often have to function under outdoor arctic conditions, low pour point and low viscosity oils are again mandatory.

The presence of water must be avoided in lubricants operating at low temperature, due to the possible formation of ice crystals causing blockages, or interfering with working clearances. Another factor which has to be considered by the oil manufacturer is that most additives show a decreased solubility in mineral oil at low temperatures. Additive precipitation is therefore more likely to occur and a careful selection has to be made.

The synthetic oils, such as the organo-silicates, are occasionally utilised in low temperature applications. However, the usage is normally confined to small specialised outlets due to their high cost. If the price was not restrictive, the organo-silicates would be ideal for many more low temperature applications, as they retain low viscosity values to very low temperatures and are still fluid at values approaching minus $50°C$.

Nuclear Radiation Environments

In nuclear engineering applications, such as power stations, the safety shielding ensures that the majority of the lubricants employed at the site are never exposed to radiation. Conventional lubricants and greases are therefore entirely satisfactory.

However, in a few isolated applications, lubricants resistant to radiation are required. Small quantities of greases are utilised in remote handling devices and other lubrication points, such as charging equipment and spent fuel systems, exposed to high levels of radiation. Conventional greases and liquid lubricants can withstand moderate doses of radiation (up to 100 Mrad) and those based on aromatic type oils, possessing ring-structured hydrocarbons, can absorb the most without ill-effe

However, exposure to excessive amounts of radiation can cause structure break-down in greases and also the deterioration of oils.

The absorption of high doses of energy, by hydrocarbons, results in chemical bond ruptures which can give rise to the formation of smaller hydrocarbon molecules. Sometimes, however, the pattern may be reversed and the activation of the hydrocarbon may result in polymerisation reactions, with an increase in the molecular size. This latter effect causes an overall thickening of the oil, in contrast to the thinning phenomenon initiated by the first effect.

In these cases, when radiation damage may be anticipated, special grades of greases and lubricants may have to be employed. The use of aromatic fractions, reinforced with special "anti-rad" additives have shown most promise. These materials have a greater potential for absorbing energy. This is, no doubt, due to the more mobile electron configurations they possess, in comparison to the more rigid electron structures of, for example, straight chain paraffinic molecules.

Other special greases may also be needed in certain instances where inert gas is utilised in the nuclear application, as the grease then has also to lubricate satisfactorily in the absence of oxygen. Under normal circumstances, the air supplies oxygen which forms an oxide layer on the metal surface. The lubrication of such a surface requires a different approach to the one which is devoid of the oxide layer. All the special grades of nuclear greases must be compatible with the many varied types of constructional materials used in the nuclear field. These materials differ from those normally encountered in conventional industrial grease lubrication.

Certain nuclear power stations have oils in the circulatory system serving the "blowers" in the gas cooling system. The oil used in this application must possess a low vapour pressure, to reduce the risk of vapour contamination of the gas coolant. The circulatory oils are not exposed to radiation, but for efficient and long service lives they require to be high quality oils, containing both oxidation and corrosion inhibitors.

Space and High Vacuum Environments

In the case of lubrication under high vacuum conditions in industry, it will be apparent that mineral oils will have a limited application, due to the vapour pressure contribution which arises from the hydrocarbons present. Specially fractionated oils are sometimes prepared to lower the overall vapour pressure, by the removal of the more volatile components.

However, in this specialised field, the silicone oils find ready outlets, due to their extremely low vapour pressure characteristics. An alternative, in some high vacuum applications, is to use solid rather than liquid lubricants. Although caution has to be exercised, as the removal of adsorbed gas films from solid lubricant surfaces can adversely affect their lubrication performance and it is unavoidable that gas desorption will take place from solid surfaces under high vacuum conditions.

In addition to the effect of a vacuum on solid lubricants, a similar phenomenon has also to be met under space environments, where the continuous high vacuum conditions have a degassing effect on all the metal components present. If the oxide layer present on two contacting surfaces becomes ruptured, then there is no oxygen present from the air to reform the oxide film. This affects the lubrication mechanism and the nascent metal surfaces will also have different reaction rates with any additives present in the lubricant. Light alloys, such as aluminium, are frequently encountered in space vehicles and although the oxide film is tenacious and normally makes the aluminium inert, an oxygen denuded surface becomes active and prone to corrosion. The use of low friction coated materials, under vacuum environments, may be encountered in certain applications. These may obviate the need to utilise a liquid lubricant.

Corrosive Environments

Many machines have to work satisfactorily under hostile corrosive environmental conditions. These may be in the chemical industry or, for example, in the paper and pulp manufacturing industries and also certain sections of the textile industry. Equipment may also have to operate near marine environments, with the presence of salt spray in the air. Industrial atmospheric pollution also plays a considerable role in the early deterioration of industrial machinery. The production of small quantities of sulphur dioxide and trioxide, by the combustion or decomposition of the sulphur compounds present in fuels and oils, may lead to the formation and condensation of acid products on metal components. The application of permanent protective coatings and paints to machine surfaces and the use of the range of temporary corrosion preventives described in chapter 5 offer some assistance in the fight against corrosion.

The initial selection, by the machine designer, of the most suitable corrosion resistant steels for the expected environment is also of paramount importance. This selection may have been limited by the mechanical properties for the steel and the anticipated working stresses. Costs would also have influenced the original decision. In certain instances, it may even be possible to replace metal bearings, for example, by chemically inert plastic types, if exceptionally corrosive products are expected to be encountered in service. In addition to the constructional materials used, good design to avoid the trapping of corrosive liquids, for example, in external crevices on machine surfaces, is also an important consideration in the prevention of corrosion.

The industrial lubricants used for the lubrication of the interior moving parts of the machines must normally be those designed and recommended for the individual applications. Some lubricants are specifically formulated to combat acidic atmospheres by the inclusion of relatively large quantities of basic corrosion inhibitors both of the liquid phase and vapour phase types. Others are designed to give efficient service under the combination of both extremely wet and corrosive conditions. In cases where the ingress of large quantities of water may have a tendency to wash the lubricant off metal surfaces, with a possible resultant lubrication failure, it has been seen that it is often advisable to use a mineral oil compounded with a fatty material. Examples may be found in drilling equipment operating under wet conditions, bearings in laundries and many marine applications. Under certain conditions, even small quantities of water may give rise to corrosion. The corrosive fretting of gears, for example, arises due to a combination of the presence of oil, water and oxygen in contact with the metal, under specific conditions of vibration and near stationary load.

The different lubricants used in various industries are discussed in detail in the relevant chapters dealing with the individual items of industrial equipment. The additives included in the products are normally satisfactory for combating the most severe corrosive conditions expected in the industries for which they were designed. Great care would have been taken by the oil manufacturer that the additives included would not be potentially corrosive to any of the metal combinations present in the system, for which they were formulated. The base oils themselves would have been selected by the manufacturer so that they would have no reaction with any corrosive chemical substances they would be expected to encounter in service. Extra care is necessary with greases, to ensure that the soaps used to thicken conventional greases are not broken down by contact with corrosive materials, such as acids, with the formation of hard abrasive solid deposits. In some such cases, it may be admissible to use a petroleum jelly rather than a standard grease. Petroleum jellies are inert to a wide range of corrosive substances.

Abrasive Environments

Dusty abrasive conditions may often be met in heavy industries such as mining, steel, brick and tile manufacture and cement manufacturing plants. Dusty abrasive conditions may also exist in many other industries due to the location of the industrial plant in desert regions. The presence of sand or dirt in an oil will result rapidly in a lubrication failure. The additional increase in the amount of wear debris formed from the abraded metal surfaces also further aggravates the problem.

In all cases, enclosed systems and good seal designs to prevent the ingress of dust are essential. Liberal applications of lubricants, at regular intervals, are also vital to prolong the life of the machinery and prevent costly break-downs under these difficult conditions. The plentiful use of lubricant aids the sealing and reduces the quantity of dust which may reach and cause abrasive wear of bearings. The very fine dust encountered in the cement industry makes a good example of an exceptionally troublesome environment. In this type of case, it is advisable to use grease rather than oil for as many bearing applications as possible. Grease acts as a better sealant than oil and reduces the amount of fine dust which reaches the bearings.

Dust, when in suspension in a grease or oil, forms an abrasive paste. It is essential to re-apply the lubricant liberally, regularly and as often as possible, not only to assist the sealing but also to have a diluent effect on any dust which has managed to gain entry. The size and number distribution of the dust particles plays a role in the potential damage it may cause on obtaining access to a piece of industrial equipment. Particles smaller than the smallest working clearances, present in the machine, will be less of a problem than abrasive particles which are larger. Although the liberal application of lubricant to interior working surfaces is essential, care must be taken to reduce the amount of oil left on the exterior surfaces of machines. Exposed metal surfaces, coated with oil, provide an excellent trap to catch dust particles.

In oil circulation systems, the importance of good filtration will be very apparent, when dirty conditions prevail. With industrial engines, the efficient filtration of the air-intake supply will be also very important to prevent the ingress of dust particles to the engine.

The storage of the lubricants and greases in abrasive and desert environments must be carefully supervised. Storage should always be under cover and any dispensing of lubricants should also be carried out under cover, because as soon as a drum or barrel is opened there is a risk of contamination. Drums should never be left open to the atmosphere and great care must be taken during the filling of a system to prevent dirt contamination taking place.

CHAPTER 9

Miscellaneous Industrial Oils

There are many miscellaneous industrial oils which are used for various purposes in different applications. Some of the more widely used of these will be discussed in the present chapter.

Industrial Engine Oils

Stationary internal combustion engines are used in a variety of industries for the generation of power. The most common type of engine utilised for this purpose is the diesel. This type of prime mover is a heat engine which converts the heat, obtained by the combustion of a fuel into mechanical energy. This energy is normal in the form of the rotary motion of an output shaft. The rotational energy may the be used to drive such items as belts, chains, gearing, ropes or pumps in a factory or works. The mechanical energy may sometimes be further converted and used for th generation of electrical power, which may be a more convenient energy form to emplo in a particular process.

In the diesel engine, air is compressed by pistons in cylinders, so that the temperature of it becomes high enough to cause the spontaneous ignition of an injected fuel. No electrical spark is needed as an ignition source. The process is therefore a compression ignition, followed by the hot gases in the cylinders expanding and pushing back the pistons, thereby creating mechanical energy. The fuel injecte into the cylinders may be diesel, gas oil or fuel oil. Industrial engines are some times designed to run on gaseous fuels, such as natural gas. Although the design i similar to the diesel engine, modifications are made so that the combustion reactio in the cylinder is normally initiated by spark ignition. It therefore differs from a true diesel.

For practical purposes, all industrial diesel engines are designed with reciprocati pistons. They work either on the two-stroke or the four-stroke principle. The oil engines can be arranged with their cylinder axes either vertical, horizontal or at an inclined angle, such as in the vee-type engine. The type favoured will normally depend upon the power, weight and size requirements for the individual industrial application.

With regard to general lubrication of the engines, the main function of the oil is to reduce friction and also to help cool the engine by the absorption of the frictional heat. A series of different viscosity level engine oils have to be available to cope with the varying design requirements of the many types of engines in service.

All the oils must possess good oxidation and thermal stabilities, to reduce the formation of sludge and deposits. This gives a long service life before it becomes necessary to have an engine oil change. Good oil filtration in the engine design is also closely associated with long oil life. The oil has the function, besides that of friction reduction, to keep the engine clean by sweeping away metal wear particles from fine clearances and from between all the surfaces in relative motion. The wear particles must be kept in suspension in the oil, together with dispersed decomposition and fuel combustion products, until they can be removed by the engine oil filtration system.

Diesel engines in industry operate under a variety of conditions. These range from low speeds at low steady outputs, through the more highly rated engines with variable outputs, to the heavy rated engines with high output duties. As the duty on an oil becomes more arduous and heavy, the necessity to have an oil with a detergent dispersant additive becomes more important. Highly rated engines tend to accelerate oil break-down and the formation of deposits. An oil with a high level of detergency keeps the pistons clean and reduces the wear rate of cylinder liners and piston rings, thereby extending their service lives. In particular, a detergent additive prevents decomposition products from being deposited on piston ring grooves, oilways and other engine parts. The temperature of the oil may affect the dispersant qualities of the oil. Good dispersion at low temperatures is more difficult to achieve but is often an essential requirement as, for example, in the crankcase lubricants of high speed diesel engines with high outputs.

There are other requirements of detergent oils, in addition to their ability to disperse degradation products. The additives incorporated in the oils must themselves be of low ash forming tendencies, in order to reduce the possibility of deposit build up on exhaust valves. The additives must also possess good anti-wear characteristics and anti-corrosion properties. The additives are always surface active in nature and are also usually of an alkaline type, to combat corrosive acidic products formed from the combustion, decomposition and chemical reaction of the sulphur compounds originally present in the fuel. These acidic products of fuel combustion will cause corrosive attacks on bearing metals, unless they are neutralised effectively and quickly.

The main difficulty, in general, with engine lubrication is encountered at start-up, before a full stable oil circulation has become established. The type and design of engine will determine the precise application method for the oil lubrication system. Priming pumps are sometimes used on very large engines, to ensure an adequate flow of lubricant is established at start-up. With regard to actual lubrication, the engines may be considered as either open or totally enclosed designs. The open design is normally used for horizontal engines. The cylinders of large engines of both horizontal and vertical types are normally lubricated by the use of mechanical lubricators. The cylinder lubricants are kept separate from the crankcase lubricant and therefore only new clean oils are ever fed to the cylinders. In the crank chambers of smaller engines operating at medium and high speeds, the cylinder lubrication relies on splash and oil mist from the crankcase. Although open horizontal engines use mechanical lubricators to supply oil to the cylinders and gudgeon pins, the main bearings often possess ring oilers. In addition, lubricant is supplied to the big end bearings by banjo ring oilers. The camshaft bearings and gears are frequently lubricated by an oil bath supply system.

In enclosed vertical engines and most modern diesel engines, the bearing lubrication is carried out by a force feed circulation system. This method ensures that an adequate supply of oil always reaches and lubricates the bearing surfaces. The oil is fed to them under pressure and the more heavily loaded bearings cannot then squeeze out the oil film which separates the surfaces.

In this summary of industrial diesel engine oils, it will have become apparent that the oils have to function under different conditions from those encountered by all the previously described industrial lubricants. Engine oils have to operate in the environment of fuel combustion products. This condition makes them unique from the other industrial lubricants.

Steam Engine Oils

Oils used for the lubrication of steam engines in industry are required to work under different environmental conditions than the oils employed for industrial diesel engines. In steam engines, combustion products are absent but there is the presence of steam. The cylinders of steam engines are not cooled as is done in the case of the diesel engines. The steam engine oils are therefore often required to operate at high temperatures, in the presence of the steam under various physical conditions. The steam condition may be that of high pressure and temperature, sometimes superheated to above $350°C$. In some engines, the steam may be dry and saturated, whilst in others it may be of low pressure and therefore of relatively low temperature, with the probability of a wet condition.

With high steam temperature engines, the cylinder lubricant is subjected to exceptionally severe carbonising conditions. The oils must be extremely stable at high temperatures. They must also be of high viscosity, so that at the high operating temperatures the oils are still viscous enough to act efficiently as lubricants. The types of oils employed for these slower speed steam engine applications are called dark cylinder oils. They are frequently compounded with fatty oils, to ensure that the possibility of water condensation on the cylinder walls does not impair their lubricating ability. However, compounded oils can only be used when any intended application for the exhaust steam is not critical with regard to the possible presence of oil as a contaminant. This can often occur due to the difficulty of oil separation from the condensate.

In higher speed steam engines, the use of slightly lower viscosity oils, or so called "bright oils", is often preferred for the cylinder lubrication, due to their ability to spread more quickly when introduced into the cylinders. If the conditions warrant it, the "bright oils" may also be compounded with fatty oils.

Air Filter Oils

Oils are frequently employed in industry to wet filter surfaces, which are used to abstract dust and dirt particles from the air. The filtered air may be intended for usage in the air conditioning system of a building. Alternatively, it may be intended for the air intake supply of an industrial engine. The reason for using an oil on a filter surface, is to facilitate the trapping of dust particles by the wetting action of the oil. The tackiness of the oil also allows the dust particles to be retained on the filter surface.

It will be mentioned later that white oils are sometimes employed in the food industry as air filter oils, especially in buildings where food is stored. In less critical industries with regard to air purity standards, conventional mineral oils are normally utilised. For different applications, various designs of air filters are available which demand different viscosity level oils. Some air filters rely on the electrostatic precipitation of the dirt particles from the air. The filter oils used are sometimes of the emulsifiable type, so that they can easily be washed off the filter surfaces when cleaning becomes necessary. Other types of air filters are arranged to rotate through an oil bath, to pick up oil on their surfaces automatically. A new oil coated filter surface can then constantly be presented to the incoming air stream. The filters are also frequently of a self-cleaning design, so that the dirt laden oil layer can be automatically scraped off the filter surface. The oils used are highly refined and possess low volatility characteristics. This reduces to a minimum the traces of hydrocarbon vapour which may possibly be removed from the filter surface by the flowing air stream. With regard to ease of wetting of dust particles, low viscosity oils will "wick" or spread over solid dirt surfaces more rapidly than higher viscosity oils. On the other hand, the lower viscosity oils will possess higher volatility characteristics and so a compromise must be made.

The viscosity of an air filter oil is also of great significance when it is used on filters connected to the air intakes of industrial engines. Large filtration panels are frequently utilised on industrial diesel engines to ensure that the intake air is free from dust. The use of clean air reduces wear on cylinders, pistons, valve seatings and lowers maintenance costs. The types of filters involved are usually of a robust nature and they are normally made of metal. They are often of a thick panel construction, with a crimped herringbone filter mesh. This mesh is designed to be wetted by the oil before the filter is put into service. The filters are especially beneficial for use in countries where dusty conditions normally prevail, or in factories with exceptionally dirty atmospheres. The greater the weight of air filter oil which can be made to adhere to the filter, the greater is the dust load which can be retained on the filter, before cleaning becomes necessary. It will be apparent that a high viscosity oil will satisfy this requirement better than a low viscosity oil. A low viscosity oil will tend to drain off the filter during its initial application and a relatively small oil film weight will be retained.

However, with regard to ease of dust surface wetting, a low viscosity oil will again be beneficial. There are therefore conflicting requirements. A high viscosity oil is desired for high dust load retention on the filter, whilst a low viscosity oil is preferable for ease of dust wetting. In order to try and satisfy these different requirements, a gelled filter oil is often recommended. This consists of a relatively low viscosity oil, which has been given an apparent high viscosity with the use of a gelling medium. The thin oil can then bleed from the gel structure to wet efficiently the dust. At the same time, the semi-solid gel structure allows a high weight of oil to be retained on the filter during the initial dipping application. Gel type filter oils are normally heated before the panels are completely immersed in them for the initial coating application.

When the oil on a panel eventually becomes saturated with dust in service, it is essential that the gelled used oil can be easily removed or laundered from the filter surface. The filters are normally cleaned on a fixed servicing basis, after being in use for a certain length of time in a particular application. The length of time between cleaning schedules, will be determined by the dust load the filter has to cope with in service. If a filter is not cleaned regularly, then the pressure drop across the filter will increase sharply. Eventually, complete filter blockage may occur. The excessive saturation of an air filter oil with dust results in a considerable increase in the apparent viscosity of the oil on the filter element. Eventually, the oil surface will become completely covered with a solid layer of adhering dust. It will then become impossible for further dust particles to be wetted by the oil bleeding from the gelled air filter oil structure. The air filter will also be much more difficult to clean if it is allowed to reach such an excessively saturated dirt laden state.

Concrete Mould Oils

Concrete mould oils are used to achieve a clean release of the formwork from the concrete cast within it. Mineral oil based products represent the most widely used class of concrete release agent in industry. However, in addition to the mineral oil based products, there are also available many synthetic chemical release agents which give entirely satisfactory performances.

There are four main types of mineral oil based mould release agents; namely, straight mineral oils, mineral oils with additives, water in oil emulsions and oil in water emulsions. They are normally applied to formwork surfaces by either brush, swab, spray or roller. The formwork materials, used for shuttering, are usually made of wood, steel or plastic. Concrete can be cast without the use of a release agent on the formwork, but the use of a release agent to prevent physical contact between the concrete and the formwork, improves the quality and appearance of the finished concrete surface. With absorbent formwork materials, it is important that sufficient mould oil is applied to ensure a continuous film is present on the shuttering surface

The main appearance faults encountered in cast concrete are blow-holes (caused by air bubbles trapped in the concrete surface), efflorescence (powdery patches which develop on the surface after weathering) and retarded hydration of the cement (discolouration often caused by absorbent formworks). The straight mineral oils are used as mould release agents when the appearance of blow-holes in the concrete, or the development of efflorescence, is not considered of importance for the end usage of the concrete. However, the use of additives in mineral oil reduces the incidence of these faults and such oils are used when the application for the concrete demands a good finish. The additives are frequently surface active agents which modify the surface tension forces present in the oil film between the mould and concrete. This allows a more rapid release of air when the concrete is cast in the mould. The concentration of additive used must be carefully controlled because its hydrophilic nature, or water soluble tendencies, may interfere with the setting of the concrete.

The use of water in oil mould emulsions gives a similar performance to the mineral oils containing additives. This is because in the emulsion, the oil forms the continuous phase and therefore is in contact with the mould and concrete rather than the water. However, emulsions have two main disadvantages. Firstly, they may cause more rusting than neat oils, when used on steel forms. Secondly, their storage stability is inferior because they are in emulsion form with a creamy nature and also their water content makes them prone to the weather conditions, especially frost The use of the reverse type of emulsion, the oil in water, is much less favoured. Water forms the continuous phase of the oil in water emulsion and therefore it can grossly interfere with the setting of the concrete, due to it mixing with the water present in the surface of the concrete. However, the use of water based release agents considerably reduces the incidence of blow-holes.

Wooden formworks are widely used as shuttering for concrete moulding but different woods vary in their water absorbencies. This can lead to hydration discolouration occurring on the concrete surface, unless the wood is saturated with oil to nullify this variation.

Steel moulds are completely impermeable to water and are therefore much better in this respect. However, they are subject to rusting which causes discolouration of the concrete. Also, steel moulds have a tendency to produce concrete surfaces with highly glazed appearances which may later develop a series of "crazing" faults. The use of suitable release agents reduces this disadvantage. Steel moulds are now frequently employed in the prefabricated house building industry and for this purpose the moulds are often steam heated. Consequently, the temperature of the mould is higher than during normal usage, so care has to be taken to select the correct viscosity for the mould oil, in order to avoid excessive oil drainage from the upper sections of the mould.

Plastics possess similar characteristics to steel when used as formwork materials. They yield concrete surfaces often with glazed appearances, which later develop "crazing" faults. The use of mould oil assists in reducing the problem but only if the surfaces of the plastic are wetted effectively by the oil. Oils have different spreading characteristics on plastics of different types, so the type of additives present and the viscosity of the oil will both be of significance in this respect.

To ensure a continuous oil film is obtained during application, it is normally advisable to use a roller or spray gun, rather than a brush or swab. With certain plastics and other special mould materials such as rubber, it may not be possible to utilise a mineral oil based product. This is because the hydrocarbons in the oil may cause swelling or other deterioration of the mould surface. Special chemical release agents may then have to be utilised.

In processes where the mould formwork is not in static contact with the cast concrete, it is again normally necessary to use a special chemical release agent. An example of this type of application is in the spinning of concrete pipes, where the mould surface and the concrete are in rapid relative motion.

If the particular mould design possesses deep valleys, or other steep recesses, it may not even be possible to use a liquid mould release agent. Liquid would tend to drain to the bottom of the recesses and gather in pools. The properties of the cast concrete would then be severely affected. In these cases, it may be necessary to cast the concrete without the assistance of a release agent.

In general rough casting on building sites, where the cheapness of the mould release agent may be of importance, the use of the oil emulsion types are often favoured. Cheaper products such as diesel, gas oil and even used engine oils have also been utilised. However, the employment of these latter materials is not recommended on health hazard grounds, due to their possible carcinogenic nature and effects on prolonged skin contact.

White Oils

The severe refining of lubricating oil fractions with concentrated sulphuric acid, produces a general class of oils known as the white oils. The sulphuric acid reacts with both the unsaturated and the aromatic hydrocarbons in the oil and these are then removed by separation in the refining process. A by-product of the refining operation is the formation of petroleum sulphonates which are recovered, due to their potential use as surface active agents. After the separation process has been completed, the white oil is neutralised and washed free of acid residues. The precise quality of this white oil will depend upon the amount of acid used in the refining process and the exact refining conditions employed. The class known by the simple name of white oils has been more highly refined than the class commonly known as the technical white oils. Both types are expensive to produce and the former class is more expensive than the latter class.

White oils are odourless, colourless and tasteless. These properties make them ideal for applications in the food, cosmetic and medicinal industries. Various viscosity grades exist for use in different types of processes. The grades have to possess good colour stabilities upon storage and exposure to light.

White oils are usually refined to meet the high purity requirements of the various pharmacopoeias, such as the British, United States of America and European specifications. These specifications lay down stringent tests to detect the presence of unwanted impurities. For example, specifications include tests to detect the presence of undesirable aromatic hydrocarbons. This is done by specifying a colour maximum in an acid char test. Polynuclear aromatics are detected by the use of ultra-violet spectrometry methods. Tests are also specified to detect the presence of sulphur compounds, measure acidity and also the ash content of the white oil.

The use of pharmacopoeia quality oils is allowed by legislation for food processing but strict controls are laid down as to the total amount of oil which is allowed to remain, as an impurity, in the food product. The requirements vary in different countries but the permissible quantities are always in small fractions of a percentage by weight of residual oil. White oils are used in the food industry as process aids in such applications as anti-stick products for coating slabs and moulds. They are also sometimes employed in minor amounts in the manufacture and preservation of certa food products.

Conventional oils and greases are utilised for the lubrication of the majority of food manufacturing and handling machinery but white oils are used in any application where it is possible for oil contamination of food to take place. The use of white oils may, in certain cases, be also relevant in the manufacture of food containers and wrappings, such as foil. The quality of any residual oil on the container will then be known to be of an acceptable standard. An alternative method is the use of natural edible products, such as vegetable or animal oils, rather than a white mineral oil based product.

In the particular case of the manufacture of aluminium strip or foil which is to be used for food container purposes, the problem is often tackled in another way. Volatile oils are used in the rolling process which will eventually be completely "flashed off" during the annealing manufacturing stage. Any additives employed in the rolling oil are also of a fairly volatile nature and they are normally derived from edible products, so that any residues which are not "flashed off" during annealing will be relatively innocuous.

White oils are sometimes used as air filter oils in the filtration systems of buildings, where food is being stored or processed. Bacteriacides and fungicides are sometimes included in the white oils for this application. This helps to ensure that the air is not only of minimal dust load but also is relatively free from harmful organisms.

The temperatures employed in the food industry are relatively low and rarely exceed $150°C$. This is because of the moisture content of the food which it is desired to retain in most food products. For processing, steam heating is most commonly utilis rather than hot oil heat; transfer heating by liquid phase is therefore fairly rare. There are a few exceptions where higher temperature processes are carried out, such as potato crisp or chip manufacture, which involve frying in vegetable oils. Anothe example of a fairly high temperature process is the deodourising of vegetable oils of the edible type. In these processes, hot mineral oil closed circuit systems are frequently used for the heating. In theory, it may be thought necessary to use a white oil as the heat transfer fluid for heating the vegetable oil, in case of leakage and possible contamination. In practice, it is found that white oils do not possess adequate thermal stabilities for this application and therefore conventional heat transfer oils are employed which are not, of course, of pharmacopoeia quality.

To avoid the possibility of contamination, the ideal design of a chip fryer, for example, would be for the vegetable frying oil to be under a slight overpressure. This would ensure, in the event of a rare leakage in the heating system, that the vegetable oil would pass into the hot mineral oil circuit, rather than the reverse case of the mineral oil leaking into the vegetable oil. As a routine precaution, stringent analytical tests are always carried out in such heating processes to ensure the vegetable oil and food products produced have not become contaminated with mineral oil. Hot oil, heat transfer systems are also employed in the food industry in "scraper surface" heat exchangers which are used in many different applications.

White oils are widely employed in the pharmaceutical, cosmetic and toilet preparation industries. The products in these industries may be intended either for interna consumption or external use. It will be obvious that the oils employed must be of medicinal pharmacopoeia white oil standards.

Technical white oils are employed in industries where a highly refined non-medicinal quality oil is required. Technical white oils are sometimes utilised for the production of certain agricultural and domestic sprays. The use of a water-white oil will be advantageous in applications where staining may be a problem. The textile industry is a good example of such an industry.

On an historical note, it is interesting to recall that white oils in the textile industry were originally introduced as a precautionary measure against skin cancer amongst textile workers. In earlier years of this century, many incidences of cancer amongst mule-spinners were diagnosed. It was recognised that the mineral oils, used at that time in the industry, possessed carcinogenic properties. Mule-spinners were especially prone because their clothing unavoidably became soaked in oil, due to the type and design of the machines they were operating.

Attempts were made to lay down specifications for oils, with a view to reducing the carcinogenic risk. The most well known of these was the Twort specification, which was introduced to control the type of oil which could be used for the lubrication of mule spindles. The Twort index was based on the density and refractive index of an oil. A relationship was established which was thought could be helpful to select non-carcinogenic oils. The use of white oils was favoured by the test relationship and in 1935 the Twort specification was replaced. Instead, a Factories Act was introduced which laid down that all lubricants used for mule spinning must be highly refined white oils. The quality of the oil was strictly defined by its colour and acid treatment.

The use of white oils in the textile industry is important, not only from the non-carcinogenicity aspect but also from the cloth staining point of view. Certain high speed sewing machines utilise technical white oils, in an attempt to avoid stain problems on the fabrics being processed. However, this is not the complete solution to the problem because even initially invisible oil stains tend to oxidise and become slightly coloured with the passage of time. The stains have a tendency also to pick up dirt from the atmosphere which further aggravates the cloth soiling problem. Various "spotting" cleaning agents are employed in the textile industry for the removal of such stains from soiled fabrics.

In addition to sewing machines and other textile machinery applications, technical white oils are also sometimes employed as carriers for additives in the preparation of certain textile fibre "spin finish" formulations. These are the chemical agents put onto the surfaces of synthetic fibres by the manufacturers to act as lubricants and to prevent static electricity build up on the fibres. In addition to this particular application, the use of water-white oils will be generally advantageous as a base for the preparation of textile fibre processing lubricants, such as the coning oils. Technical white oils are sometimes used as an integral component of certain synthetic textile fibres and the oils are incorporated into the fibres during their manufacture. The oils function as internal fibre lubricants.

Petrolatums

Petroleum jellies, or petrolatums, find specialised applications in industry. Petrolatums are mixtures of mineral waxes and oils. They are stabilized in such a fashion that the oil appears to form the internal phase, whilst the wax compounds seem to constitute the external phase. The petrolatums are fibrous or grease-like in nature. They possess discrete drop points and also penetration values. They are prepared by dewaxing the residues obtained from paraffinic crude oil fractionation. The crude petrolatum obtained in this way is then further purified to varying degrees, which depend upon the intended end use of the petrolatum.

Some petrolatums are purified to such an extent that they meet the pharmacopoeia petroleum jelly grade requirements of various countries, such as Britain and the

United States of America. There are two such grades, one for white and the other for yellow petrolatum. These can be employed for the production of various medicinal products, such as ointments and creams for external use on the skin.

Less refined petrolatums are normally green in colour and are limited to industrial technical applications. Typical examples are thick film temporary corrosion preventives and inner hemp core wire rope lubricants.

Slack Waxes and Refined Waxes

After the subject of petrolatums, it is opportune to mention the so-called slack waxes and also the more refined waxes. Petroleum waxes are obtained by separation from lubricating and gas oil fractions in refinery processing. Slack waxes consist of mixtures of mineral oil and wax in various proportions, but they do not possess the structure and fibrous nature of the petrolatums. They are also less refined and may contain up to sixty per cent of oil. Slack waxes find many applications in industry, such as in rope lubrication and as leather dressing compounds. They are also sometimes used as softeners and plasticizers in the compounding of certain rubbers.

Refined petroleum waxes consist of mixtures of highly refined paraffinic hydrocarbons which possess relatively high melting points. The waxes are therefore solid at normal temperatures. There are two main classes, the paraffin waxes and the microcrystalline waxes. The various grades will possess different melting points and flexibility characteristics. Both types find many uses in industry. These range from textile fibre lubrication, candle making, polish production, the waterproofing of materials to the manufacture of cosmetic preparations and the production of waxed paper. The use of waxed paper and cartons are frequently encountered in the food container industry. It will therefore be apparent that the wax employed for this purpose must be of the highest quality refined standard.

Petroleum Extracts and Bitumens

A class of oils known as the petroleum extracts are much used in industry in a large number and variety of processing operations. They are normally of an aromatic nature. The many individual aromatic hydrocarbons vary in molecular weight but are usually in a fairly high boiling point range. They are also frequently in a complex mixture with similar molecular weight range naphthenic hydrocarbons. The types of hydrocarbons present result in the petroleum extracts having relatively low viscosity indices. Care must be taken in the general handling of petroleum extracts to avoid excessive skin contact or inhalation of mists. This is because they are likely to contain condensed aromatic compounds which may be carcinogenic in nature. It has already been mentioned that the petroleum extracts are produced from lubricating oil fractions by solvent extraction. A variety of viscosity ranges are normally available for use in a range of process applications.

One of the main uses for petroleum extracts in industry is as extenders and plasticizers for rubbers and plastics. In the rubber industry, they are used as extenders and process assistants for both natural and synthetic rubber mixes. They are normally mixed into the rubber at the same time as the carbon black, which is frequently employed in tyre manufacture as a reinforcing compound and filler. Carbon black is a particular form of carbon possessing a very small particle size. It is prepared from a petroleum feedstock which may, in fact, be a petroleum extract.

Bitumens derived from petroleum are very complicated mixtures of asphaltic residues and heavy oils. However, the types of hydrocarbons are mainly thought to be relatively harmless to man with regard to health hazards. Bitumens may roughly be divided into two main classes known as the penetration and the oxidised grades. The former class find their main application in road surfacing duties. The oxidised types find many industrial applications in addition to those already mentioned, such as cable insulants, battery fabrication and bituminous paint manufacture. They are also often utilised as the basis for motor car chassis protective compounds or undersealants. In this application, the underseal layer must be thick enough to withstand mechanical damage. Rubberised compounds are frequently also incorporated to improve the characteristics and reduce road noise, by acting as anti-drumming agents.

CHAPTER 10

Care of Industrial Oils

Storage

Care in the storage of industrial oils is required, to avoid any changes taking place in their compositions before they are used. In the case of products containing volatile solvents, such as temporary corrosion preventives, soluble oils, wire rope and open gear lubricants, it is essential not to expose them to heat. Excessive exposure to a hot sun, or storage near steam pipes in a factory, results not only in solvent loss but also in a potential fire hazard, due to the build-up of a hydrocarbon vapour in the atmosphere.

In the case of products containing water, such as fire-resistant hydraulic fluids, concrete mould oil emulsions and cutting fluids, it is equally important not to expose them to cold or frosty conditions, which could result in ice separation. In addition, fatty components present in such products as temporary corrosion preventives and cutting oils, show a tendency to come out of solution under cold conditions

With all products, it is essential to store them under dry conditions, in order to avoid the pick up of atmospheric moisture. Most neat industrial oils have a moisture content of less than 50 parts per million and electrical oils have even less. For this reason, electrical oils are usually stored in special epoxy resin coated tanks, fitted with silica gel air vents to avoid moisture contamination. The other main contaminant which must be avoided during the storage of industrial lubricants is dirt. This causes havoc with the use of such products as greases and also hydraulic oils, when operating in systems containing very fine clearances. It is therefore important to use a clean storage area and to keep dirt from the tops of containers and drums.

Heating

In the handling of non-solvent grades of industrial oils, it is sometimes necessary to heat them, in order to reduce their viscosities. In general, the susceptibility of the higher viscosity oils to thermal cracking is greater than the lower viscosity ones. (This is why the use of low viscosity oils is favoured in heat transfer oil systems rather than heavier oils). Unfortunately, in the handling of products, it is normally the higher viscosity ones which require heating. In order to ensure that no thermal damage is done, it is advisable not to heat an industrial oil above an approximate bulk temperature of 75°C. With oils containing extreme pressure agents, lower temperatures should preferably be employed to ensure that no additive breakdown occurs.

In practice, industrial oils are normally heated by either the use of steam coils or electric immersion heaters; hot oil heat transfer systems are also sometimes employed. Although the bulk temperature of the product may be kept below 75°C, care must be taken not to allow local overheating to occur at the heater surfaces, which could still result in some thermal cracking. It is therefore advisable, with steam heating, to maintain a maximum saturated steam temperature of only 160°C in the coils. Heat will then be transferred to the product at a sufficient rate to allow it to reach the maximum bulk temperature of 75°C, without overheating occurring. In the case of oils containing extreme pressure agents, the maximum steam temperature in the coils should be approximately 130°C, instead of the 160°C suggested for the non-additive oils. Similar temperature levels are also advised when a hot oil, heat transfer system is utilised instead of steam for the heating process.

When electric immersion heaters are used, careful control must also be maintained of the maximum heat flux at the heater surface. With high viscosity non-additive oils, it is advisable to limit the maximum flux to approximately 12 kilowatts per square metre of heater surface. With lower viscosity products, this value can be doubled because of their greater thermal stabilities. In the case of high viscosity extreme pressure oils, the maximum flux should be maintained below approximately 8 kilowatts per square metre of heater surface but for lower viscosity products double this flux can again be used.

In-Service Care

The method and frequency of application of a lubricant to the actual points needing lubrication in a plant will depend upon the particular design. It may be by hand lubrication, mechanical lubricators, micro-fog applications or, in the case of large systems, by piped distribution from centralised supply tanks. The latter application method, as mentioned earlier, is usually reserved for large installations such as steelworks, where both oil and grease are normally supplied to gears and bearings by pumping through a pipeline from a cellar store. In large workshops, cutting fluids are also frequently piped to the many individual machines from a centralised supply tank. In the case of micro-fog lubrication systems, the lubricant is designed to be applied by oil mist techniques. This means the lubricant is carried in mist form, in admixture with air, to the point requiring lubrication, where it is then reclassified into the liquid form to carry out its duty as a lubricant.

Care is often taken to prolong the service lives of industrial oils and treatment units are frequently built in, as integral parts of the plant or equipment design. These may range from simple coarse filters, as used in heat transfer oil systems, to the fine filtration methods used in aluminium rolling mills to remove debris of colloidal dimensions. Fine filtration removes particles of dimensions less than one micron (0.001 mm) and the filter medium is normally an activated earth. Activated earth filters also remove oxidation products from oils circulated through them. However, care must be exercised in their general use because they can also adsorb additives from oils. It may not always be possible, or practical, with fine filtration to utilise a full flow of the oil through the filter, and in these cases, by-pass filters are used. These filter only a certain percentage of the total fluid at any time and they are much used in the larger industrial systems.

In these systems, a proportion of the oil in circulation is therefore continuously withdrawn, purified and then returned to the main system. The fine filters are especially liable to block, if kept in service for too long. A gradual blockage prevents sufficient oil flow through the filter and a build-up is particularly noticeable with oils containing surface active or detergent type additives. With normal coarse mechanical filters, the surface active additives hold colloidal carbon and degradation products in suspension in the oil and the suspension can pass through the filter. With fine filters, this is not the case and the suspended degradation products are retained by the filter and eventually completely coat the surface layers.

The cleaning of the surfaces of the earth filter is carried out by the process known as back washing, which is done at intervals. This process removes the coated earth layers from the filter and carries away the impurities with it. New earth powder then has to be reintroduced into the filter shell. This can be done by adding the earth powders to some of the oil and flushing it through the filter, where the powde is retained on the filter elements.

The earth powders employed in the filters are frequently mixtures of diatomaceous si and fuller's earth. They have normally been activated to increase their absorption capacities. The fuller's earth retains colloidal sized particles, whilst the diatomaceous silica retains the larger size solids. The powders in the filter eleme are normally tightly encased within a wire mesh. A series of such elements is built up into the form of a cylindrical shaped shell.

In general industrial equipment, wear debris of less than one micron in size causes no harm and therefore there is no need to use a fine filtration method. (It is used in the case of aluminium rolling, for example, because of the prime importance of th surface finish and appearance of the rolled aluminium strip). For general industria systems, edge-type filters employing paper discs, are entirely satisfactory for the removal of particles down to a size of one micron.

Settling tanks, together with centrifugal separation methods, are also frequently employed in various industries to separate debris, of particles sizes down to one micron, from oil. Magnetic filtration methods remove ferrous swarf efficiently and are often used for this in the treatment of used cutting fluids. Many industrial plants use combinations of the various filtration and separation methods, in order to maintain oils in good condition during prolonged service lives.

It should be emphasized that all the methods just described are used to maintain an oil in good condition during service. They are not intended to overcome the effect of introducing an oil initially into a dirty plant. All systems must be effectively cleaned before filling with oil. This may involve the flushing of a system with a low viscosity oil, or sometimes flushing with the oil to be used in the system. The flushing oil is discharged after effective cleaning of the system has been carried out with it. In the particular case of machine tools, the presence of dirty sumps can lead to the immediate bacterial infection of a new soluble oil emulsion. It is therefore good practice to clean and also disinfect machine tool systems, to ensure that no residual bacteria have been left, in stagnant pockets of previously used emulsions which could not be effectively drained from the system.

In the general use of industrial oils, care must also be taken to protect plant operators from any potential health hazards. With regard to inhalation, oil mists in workshops or plants should be kept below a value of 5 milligrams of oil per cubic metre of air. The size distribution of the oil drops in the mist is also of importance, as it affects the amount of inhaled oil mist which can be immediately exhaled and therefore not retained in the respiratory system for any length of time.

In certain industries, such as in cutting fluid workshops, there is a greater risk of skin contact occurring with the oil being used than when, for example, oil is employed and contained in an enclosed plant. Prolonged skin contact with mineral oil sometimes leads to skin disorders. Good hygiene standards prevent health risks and gloves, with protective clothing, should always be worn.

Disposal

With regard to the disposal of used neat industrial oils, several methods are utilised. Waste contractors are available to collect used oils and dispose of them in approved ways. Alternatively, it may sometimes be possible to blend certain industrial oils in small concentrations into fuel oils, which can then be burnt in conventional furnace equipment. Special burners are also available which have been designed for the combustion of waste lubricating oil.

Soluble cutting oil emulsions, for disposal, are treated in special chemical treatment plants to split the emulsion into an oil and a water phase. The oil may then be burnt as a waste lubricating oil. The water phase may require further treatment to make it an acceptable effluent, for disposal into sewage systems. This is especially so when substantial quantities of biocides have been added to the emulsion. Gross dilution with further water may be essential, to ensure that the biological purification process in the sewage works is not affected by the effluent.

In the case of fire-resistant products, such as hydraulic fluids, the normal practice for disposal is collection by a waste contractor. It is not possible to burn such products because of their inherent nature, which is designed to prevent combustion. The water-in-oil emulsion fire-resistant type hydraulic fluids are very difficult to split into two phases, as can be done relatively easily with the oil-in-water emulsion type soluble cutting fluids. It is therefore necessary to have a contractor to collect and dispose of such products.

Safety in Handling

In the general handling and draining of systems containing mineral oil, care must always be taken to ensure a safe procedure is adopted. Used oils, in particular, can possess much lower flash points and higher volatility characteristics then new oils. Mineral oils consist of hydrocarbons and therefore, if a critical concentration of hydrocarbon vapour or mist is allowed to build-up in contact with air, then a potential fire or explosion hazard will exist. The conditions are then right for an ignition source to trigger off a combustion reaction. The critical concentration limits of hydrocarbon vapour, or mist, in air will depend upon many factors, including the temperature and pressure conditions, together with the dimensions and nature of the containing vessel. The safest procedure to follow is never to drain hot oil into a tank or collection vessel but always allow the oil first to cool to room temperature. The presence of possible ignition sources in the immediate area must also be strictly avoided. It must additionally be realised that a potential fire or explosion hazard still exists in the vessel, or plant, just emptied. As the oil is drained from a system, air will normally enter it, in the place of the oil. It is, therefore, possible for an apparently empty system to contain a critical concentration of hydrocarbon vapour and air. In certain instances, such as the draining of heat transfer oil systems, it is prudent to allow an inert gas, such as nitrogen, to purge the system and take the place of the drained oil.

In the case of water based industrial oils there is, of course, little risk of fire or explosion. However, the water content of the fluids during service must be monitored and any losses, due to evaporation, replaced. This is of prime importance, with regard to safety, for water based fire-resistant hydraulic fluids. A fall-off in water content would obviously increase the fire risk. Eventually, if all the water was allowed to evaporate by neglect, then the fire-resistant properties of the fluid would be entirely lost. A dangerous situation would then arise; especially, for example, if a hydraulic pipe should fracture and a spray of fluid was ejected under pressure.

Water Based Products

When glycol and water mixtures are used for heat treatment purposes, the continuous quenching of hot steel results in increased evaporative losses. The water content of the quenching fluid will again have to be checked at intervals and the losses replaced. A similar situation arises with metal rolling oil emulsions and also soluble oil emulsions used for metal cutting. The normal procedure for checking the strengths of these emulsions is to "split" them in the laboratory, with the use of an acid or salt solution. The amount of oil separated can then be measured and therefore the strength of the emulsion calculated. When the emulsion cannot be readily "split", then the strength can be determined by the measurement of the refractive index of the emulsion. The refractive index is a function of the density of the emulsion and therefore its dilution, which can then be obtained by the use of a suitable calibration curve. This curve is previously prepared by measuring the refractive indices of known strengths of the emulsion.

The "in-service" care of an emulsion is of importance but also of equal importance is the correct mixing procedure to prepare the emulsion before service use commences. In the case of soluble oil emulsions, which are of the oil-in-water type, it is essential that the oil is always added to the water, with stirring. The water should never be added to the oil. For the preparation of small batches of soluble oil emulsions, up to about 100 gallons in size, hand stirring with a paddle will normally be adequate. This is done as the oil is added, for example, to the water in a tank. For the preparation of larger batches, it is advisable to use a commercial mixing device, such as a dispersator or an injector type, which will give improved agitation. The oil is broken down into smaller sized droplets much more quickly, when this type of mixing method is employed.

The quality of the water, used to prepare the emulsion, is of importance for two main reasons. The first is connected with the water hardness. Extremely hard water may adversely affect the emulsion stability and it is therefore advisable to check the hardness before the preparation of a large batch of emulsion. On the other hand extremely soft waters may sometimes cause excessive foaming of the emulsion during service use. A water of average hardness is therefore normally preferred. This is one with a total hardness in the approximate range of 50 to 300 parts per million of equivalent calcium carbonate.

Microbial control. The second reason for the importance of water quality is the bacterial aspect. The use of a polluted water must be avoided, because of its potential detrimental effect on emulsion stability. This is in addition to the obvious objection on hygiene grounds. As mentioned earlier, bacterial populations can multiply rapidly in soluble oil emulsions and then can often degrade the emulsifiers. Bacteria can be introduced into emulsions in ways other than the water used for their preparation. Air-borne dust and extraneous matter are typical examples. "Tramp oil" may contaminate the emulsion from leaks in bearings and form a thin oil layer on the top of the emulsion, thus encouraging anaerobic bacterial growth. It is therefore advisable to monitor the bacterial population of an emulsion during its service use, so that appropriate action can be taken to prevent gross deterioration.

In addition to the chemical methods, such as the use of biocides described earlier for microbial control of emulsions and water based coolants, there are also several other methods which have shown promise. At the moment, they are not widely utilised in practice but hold promise for a wider acceptance in the future. The removal of micro-organisms by gravity methods, centrifuging and sterilization by heat are typical examples. Ultrasonic treatment and irradiation processes also show potential as control methods. The removal of micro-organisms by filtration is also of possib interest for the treatment of industrial coolants. A filtration method based on the use of ceramic filters, impregnated with a relatively insoluble toxic product such as silver, has already been in use for many years to treat polluted water and make it suitable for drinking purposes. Such filters possess extremely small pore sizes, so that they are capable of trapping even the smallest size bacteria. The use of ozone for the microbial control of industrial coolants is now also a practicable proposition. Ozone has been proposed and also utilised on a limited scale for many years for the treatment of drinking water. All these methods hold promise for the successful microbial control of industrial coolants and therefore the prolonged service lives for the water based products.

Synthetic Oils

In general industrial lubrication, it is never advisable to mix different oil grades together unless their compatibility has first been verified with the supplier. In particular, synthetic oils should never be mixed with mineral oils. If it is desired to change from a mineral oil to a synthetic fluid, then it will be necessary to clean the system thoroughly first and also change the seals and any painted surfaces which may be present. Seals and paints which are compatible with mineral oils are not usually compatible with synthetic oils, although the emulsion type synthetic fluid is sometimes an exception.

Oil Reclaimation

Earlier in the chapter, methods were described which are frequently employed to prolong the lives of neat industrial mineral oils during actual service. These included separation and filtration units built in as integral parts of the plant or equipment design. Similar methods are also utilized for the reclamation of used oils after they have been completely drained from systems. Reclaimation is the large scale batch purification of collected used oil.

Reclaimation saves money, when it can be done, because it avoids the need for scrapping and disposing of mineral oil which can possibly be later utilized for other purposes. It may be possible to use a reclaimed oil in its original application but usually it is the best practice to employ it in a less arduous one. Reclaimation is normally only a viable proposition when large quantities of lubricating oil are involved. Doubts have recently been cast on the advisability of the practice, due to possible health hazards. The repeated use of mineral oil, especially under arduous conditions, may lead to the possible accumulation of trace undesirable carcinogenic components in the oil. These components, such as the polynuclear aromatics can sometimes, as previously mentioned, be responsible for the causing of skin cancer. This is particularly so when repeated skin contact with the oil is maintained over many years, due to bad personal hygiene or the absence of protective clothing. However, in many industrial applications, for example, where the oil is enclosed, there is little risk of prolonged skin contact. In these cases, it may be justifiable to consider the viability of oil reclamation.

The main contaminants in a used oil are usually solids, water and oil degradation products. The reclaimation of an oil is designed to remove as much and as many of these as possible. The main solids are dirt and wear debris of various particle sizes. Carbon and solid sludges from oil degradation may also be present. Solids are relatively easy to remove from a lubricating oil. The larger particle sized components can usually be removed by allowing the used oil to reside in a settling tank for a period of time, several days if possible. The process of gravity solid separation can be accelerated by heating the oil to reduce its viscosity. The maximum temperature employed should not exceed $70^{\circ}C$, to avoid the possibility of any further degradation occurring. The clean oil is piped away from the solids when the separation process has been completed. If the oil was also grossly contaminated with water, then the majority of this will also be removed in the settling process.

The centrifuging of oil will accelerate the effect of gravity separation and can be employed advantageously for the removal of the remaining water contamination and finer solid particles. Most of these will possess higher specific gravities than the oil and can therefore be satisfactorily removed by the use of a centrifuge. The dry centrifuging of oils will not remove any conventional additives from the oil because they are in molecular solution with the oil. However, sometimes the technique of wet centrifuging is employed, in which a controlled amount of water is fed into the centrifuge with the oil. In this case, there is a danger that some of the oil additives may be removed, so care must be taken. It is perhaps advisable to reserve the use of wet centrifuging to non-additive oils, unless it is possible to re-dose the oil, if necessary, with more of the same additive at a later stage. However, this will depend upon the specific circumstances because it may not be intended to use the reclaimed oil in its original application.

The idea of wet centrifuging is that the water wash removes some of the solids quick from the centrifuge. The centrifuge will then require less maintenance and cleaning which is a great practical advantage. Water washing will also remove some of the more water soluble oxidation products from the oil. The acidic content of the oil may thereby be reduced. It is important that the wet centrifuging technique is only applied to oils possessing good demulsification characteristics. Otherwise, troublesome emulsions may be formed.

Oxidation products can also be removed by the use of activated earth filters. However, as mentioned earlier, care must be taken because certain conventional additives may also be removed from the oil. Solid particles which were not removed in the separation, or centrifuging process, can be separated by fine mechanical filtration methods. These filters, unlike the activated earth ones, do not remove conventional additives from oils. This is because the additives are in molecular solution and there is no surface active area present in the filter to cause additive adsorption.

Whilst most of the solids, water and degradation products can be removed from an oil by the various reclaimation processes described, it is inevitable that some traces, particularly of degradation products will still remain. These may be polymerised hydrocarbons which may cause the oil to possess a higher viscosity than when in its original unused state. In certain cases, there may be a reduction in the viscosity. This happens when thermal degradation of the oil has yielded a significant increase in the concentration of the lower molecular weight hydrocarbon In this case, the flash point of the oil may also be much lower than it was originally. It is therefore advisable, with a reclaimed oil, to check the physical characteristics in order to ascertain the degree of change from the original unused oil. It may be possible to restore the oil to its original viscosity by blending with an appropriate oil of a higher or lower viscosity. Chemical changes in the oil are more difficult to detect and detailed knowledge of the additives present and their concentration are essential, although this type of information is usually not available to the oil user. Therefore, this aspect can only be investigated by the oil supplier.

Sampling

In the care of industrial oils, we have seen it is often essential to monitor the quality of the oil in service. Some of the tests carried out to determine the relevant physical and chemical characteristics of the oil will be described in the next chapter. Before discussing these, it must be emphasised how important it is that the oil sample tested must be exactly representative of that taken from the system. To achieve this, correct sampling must be carried out to ensure that the oil undergoes no change in composition, before it is tested. Changes can occur, for example, due to the loss of volatile fractions from the oil or by the ingress of contaminant during the sampling procedure.

To avoid the loss of volatile components, samples should always be stored and transported in corked, or stoppered, glass bottles. They must be protected from extremes of temperature and also exposure to light. The bottles must be completely free of contaminants before the oil samples are introduced into them. In particular, any traces of solvents used to clean the bottles must be removed. The corks or stoppers for the bottles should either be new or free from contaminants. In the special case of samples for microbiological examination, it is essential to use properly designed sterile containers. For samples of grease, the most convenient method is the use of clean tin containers with large lids. Bottles are usually impracticable for grease samples because of their relatively small neck sizes. Greases should not be wrapped in materials which are absorbent, because some of the oil in the grease is likely to be removed. The composition of the grease will therefore alter before the sample can be tested.

The type of sample required from an industrial system will vary greatly depending upon the actual problem and application. A top, middle or bottom sample may sometimes be required from a particular system, rather than an average representative sample. Bottom samples are especially important in the case of water contamination and sludging. If taps, or other convenient methods, are not available for drawing samples from a system, then the most practical method is to employ a long open ended tube. The tube can be inserted into the oil through any convenient point which can be opened in the system. The bottom of the tube is preferably constricted or drawn.

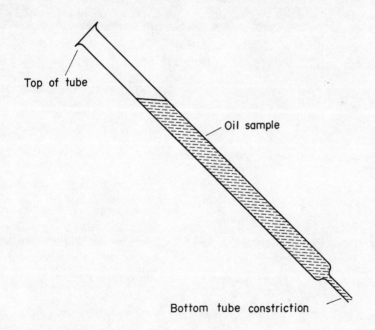

Fig. 10.1. Oil sampling tube.

When removing a sample from the system, a thumb is placed over the top of the tube. The sample in the tube is then allowed to run into the sample bottle, by removing the thumb from the top of the tube. If a bottom sample is required, rather than one representing the full depth of the oil to which the tube is immersed, then a thumb should be placed over the top of the tube before it is immersed in the oil. Oil will not enter the tube until the thumb is removed from the top, due to the air pressure in the tube. It is possible in this way to obtain a sample from any depth in the oil by immersing the sampling tube to the desired level. If a thumb is not placed over the top of the tube, then it is filled continuously as the tube is lowered in the oil. Oil sampling tubes vary in length and shorter ones can be used to sample shallow depths of oil (Fig. 10.1). The use of a sampling tube is preferable to the employment of a scoop fitted with a long handle to take an oil sample. It is not possible with this latter method to take a middle or bottom sample. The use of a bottle tied to a piece of string should never be employed in case of accident and the loss, or breakage, of the bottle in the system.

When samples are sent to a laboratory for testing, they must be clearly labelled with regard to their contents and also the precise point from which they were drawn in the system. The date the sample was taken should also be recorded. The size of sample required will depend upon the particular tests envisaged. However, an average size of 400 to 500 millilitres will be sufficient for most cases. If a sample is required from a drum or storage barrel, then great care should be taken in cleaning the lid or cap before opening to obtain the sample. Again the most convenient method will be the use of a sampling tube.

CHAPTER 11

Industrial Oil Testing

PHYSICAL AND CHEMICAL LABORATORY TESTING

The initial selection of an industrial oil for an individual application, is usually made by examining the physical and chemical characteristics of the proposed oil. The tests used to determine these characteristics are therefore very important, as they give an indication of the oil quality.

Viscosity is normally the first consideration and this is measured in an apparatus called a viscometer. From viscosity measurements determined at different temperatures, the arbitrary term known as the viscosity index can be calculated. This gives a measure of the variation in viscosity with temperature change.

The acidity of an oil is frequently determined and expressed in terms of a neutralization value. This value denotes the amount of alkali required to neutralize the acidic products in a specified weight of oil. An unused non-additive oil has a very low value but the presence of certain additives may give rise to a slightly higher value. The neutralization value is most useful in assessing the extent of deterioration of used oils in service. The formation of acidity is usually caused by the oxidative degradation of the oil. Acidity values, together with other data, assists in deciding when an oil should be changed.

A test of similar nature to the acidity test (but performed in a different manner) is carried out to measure the saponification value of an oil. This test is only relevant when a fatty material is present in the oil and the test is therefore used to indicate the presence of saponifiable material. In the test, the amount of alkali is determined which reacts by saponification with any fat present. It is important to distinguish between this test and the neutralization value test. The neutralization value gives a measure of organic acidity in the oil, whilst the saponification value is related to the presence of fatty material.

Fire, flash points and auto-ignition temperature give an indication of the maximum temperature at which an oil can be used with safety. All these tests are related to the volatility of the oil, as are also the vapour pressure and the boiling range of the oil. In general terms, the higher the viscosity, the higher will be the fire point of the oil.

Demulsification tests give an indication of the ability of an oil to separate from an emulsion or an oil water mixture. The results are expressed in terms of the time (in seconds) required for water separation to occur under standard conditions. These tests are particularly important with circulation oils and strict control tests are laid down by industry.

Standard laboratory procedures are also available to check the foaming stabilities and tendencies of oils. Standard tests are also used to determine such characteristics as the pour points, sulphur contents and specific gravities of oils.

The cloud point of an oil is determined in the same apparatus as used for the pour point. However, the two points are different and denote separate aspects. The cloud point is the temperature at which wax crystals first begin to separate, when the oil is cooled under prescribed conditions. The cloud point is not connected with any movement of the oil in the test. In contrast, the pour point denotes the lowest temperature at which the oil will continue to flow under the specified conditions of the pour point test.

The total sulphur content of an industrial oil is not normally of such a great significance, as it is in the case of a hydrocarbon fuel. However, in certain specialised applications, such as heat treatment oils, it may have an importance. For example, the oil fumes arising from a quenching bath may sometimes accidently be sucked into the furnace and come into contact with the Nimonic heating elements. In this case, corrosion may occur if the total sulphur content of the oil was not of a relatively low value.

The specific gravities, or relative densities of oils, are often measured because the values obtained are utilized, with other physical data, to assist identification of the predominant types of hydrocarbons present. For example, paraffinic oils will normally possess lower specific gravity values than naphthenic oils of comparable viscosity.

There are several standard tests to assess the potential deposit forming tendencies of lubricating oils. The ash content, after the incineration of the oil, provides a guide to non-volatile components, such as metals which may have originated from additives or wear debris. Carbon residue tests, such as the Conradson and Ramsbott are sometimes used to predict the potential carbon forming tendencies of oils. However, these tests are really intended for the assessment of fuels rather than lubricants. In the tests, carried out under different specified conditions, a quantity of the oil is heated until all the volatile matter has disappeared. The residue is then weighed and the result is quoted as a weight percentage of the quantity of oil tested.

The asphaltene content of an oil is also often determined to give additional information on the potential for deposit formation. Asphaltenes are complex high molecular weight hydrocarbon compounds which are present in oils. They are normally arbitrarily classified as the wax free constituents which are found to be insoluble in a paraffinic solvent, normal heptane but are soluble in an aromatic solvent, benzene. It is possible for an aromatic oil to precipitate deposits, if it is mixed with a highly paraffinic oil in service.

The aniline point of an oil is sometimes determined to give an indication of its aromaticity and its likely swelling effect on natural rubber. This is of obvious importance if any seals, or other components, in an industrial oil system are made of natural rubber. The aniline point of an oil is the lowest temperature at which it remains completely soluble with an equal volume of the aromatic basic compound aniline, under the prescribed test conditions. Below the critical aniline point temperature, shaking the mixture merely produces a cloudy emulsion and not a solution

The iodine value of a lubricating oil is determined with a view to indicating the quantity of unsaturated hydrocarbons present. It is also used in the identification of fats which possess distinctive iodine values, due to their various degrees of unsaturation. Always these hydrocarbons are more prone to oxidation than saturated paraffinic hydrocarbons. The test is also occasionally used to give a rough estimate of the possible oxidation stability of a mineral oil. However, the results of such a simple test cannot be used solely, or reliably, as a true indication of oxidation stability. The bromine number is also sometimes utilized to assess the content of unsaturated materials present in an oil. This test is often employed for the assessment of aluminium rolling oils, where the presence of unsaturates is particularly undesirable due to them possibly causing staining of the metal.

In the iodine value and the bromine number tests, measurements are made of the quantity of the individual halogen which reacts with a known weight of the oil. The iodine and bromine attack and try to saturate any unsaturated linkages which may be present in the sample. In the case of oils and fats, these linkages are mainly olefinic in nature. The presence of unsaturated linkages in a product can only suggest that it may be susceptible to oxidation in service. As indicated earlier, the amount or percentage of unsaturated material present cannot be used to draw a firm conclusion regarding the potential oxidation stability of the oil. This is because the stability will also be greatly affected by other factors, such as the presence or absence of natural or added oxidation inhibitors. It is difficult to detect the presence of oxidation inhibitors by simple chemical tests, although it is possible to do so by the employment of spectroscopic and other more sophisticated techniques.

Oxidation reactions in mineral oils proceed in very complicated ways and depend on many factors. It is therefore difficult to lay down a standard oxidation test, to assess the stability of an oil. However, various closely controlled tests are available which are used to give a general assessment of the expected oxidation stability of an oil in service.

For example, with circulation oils such as steam turbine oils, a standard size of oil sample is subjected to oxidizing conditions; by passing air or oxygen through it in the presence of a metallic catalyst, at a specified temperature. The tendency of the oil to oxidize is measured in terms of the acidity which develops during the test. Performance standards are usually given either in the time (in hours) to reach a specified acid value, or the acidity value attained after a certain number of hours. Oxidation tests are carried out in a different way to assess the stabilities of such products as greases. In this case, the assessment is made by determining the oxygen pressure drop which occurs in an enclosed bomb, containing the grease under specified conditions.

The environments encountered by oils and greases in service, with respect to their ability to prevent rusting or corrosion, are again difficult to reproduce in the laboratory. However, various standard tests are available and are carried out under specified conditions, with a view to giving some assessment of the possible performance of the oil in its various service applications. Humidity cabinet and salt spray tests are frequently carried out on combinations of metallic components, under a variety of conditions, when coated with the oil under test.

With regard to the general testing of greases, it is necessary to use tests designed specifically for grease assessment. This has just been referred to in the oxidation stability test and it is because greases possess a different nature to oils. Grease consistency is measured in terms of a scale based on the penetration of a standard metal cone into the grease. The depth of penetration is expressed in tenths of millimetres. In the test, the load applied to the cone is constant and the time allowed for penetration to occur is standardised. From measured penetration data, greases are given a consistency value which lies on a devised arbitrary scale, normally ranging from 0 (for soft greases) to 5 (for stiff greases). The penetration tests are usually made on a grease in both the "unworked" and the "worked" state. Standard laboratory grease workers are available to work greases and penetration values are usually determined after working in the apparatus for 60 or 100,000 strokes. The shear stability of a grease can be assessed by the difference between the worked 60 stroke penetration and the worked 100,000 stroke penetration.

The drop point is another important physical characteristic of a grease. It is determined by heating the grease and observing the transition temperature at which the semi-solid state changes to a liquid state. The drop point of a grease gives an indication of the maximum temperature at which the grease can be used in service. It should be noted that with metallic soap thickened greases, the maximum temperature at which the grease will give a satisfactory performance is some way below the drop point.

Cutting oils are sometimes assessed by laboratory tests to determine their chemical activities. In particular, neat extreme pressure cutting oils are tested to assess their potential for releasing active sulphur and chlorine. The test is carried out by heating the oil in the presence of copper at various temperatures (for example $100^\circ C$ and $200^\circ C$), under specified conditions. Any sulphur or chlorine released by the cutting oil reacts with the copper and the precise amount can be determined by chemical analysis. Active oils release more sulphur or chlorine than inactive oils. It is sometimes advantageous for a cutting oil to be inactive at the lower temperature levels, to avoid possible staining problems with machine tools and components. Activity may therefore only be required at the temperature levels reached in the cutting zone, where the extreme pressure conditions exist. It is for this reason that sulphur and chlorine release values are determined at different temperature levels.

Bacteriological tests can be carried out on soluble oil emulsions and water based fluids. A quick qualitative assessment of the bacterial contamination can be made by the use of an indicator paper, which undergoes a colour change when spotted with an infected emulsion. A quantitative assessment is best made in the laboratory by the use of standard culture bacteriological counting techniques.

For the assessment of quenching oils, two main types of laboratory tests are utilized. The first evaluates the thermal or cooling capacity of the oil, the second involves the hardenability testing of steel specimens after being quenched in the oil.

The first type of test is frequently carried out by measuring the rate of cooling of a silver ball when quenched in the oil. A thermocouple is placed in the centre of the ball to obtain the cooling curve (Fig. 11.1). Silver is used as the test specimen because it undergoes no phase changes or thermal transformations during its cooling process. With a steel specimen, several phase changes and temperature arrests occur during cooling. Also, silver shows a freedom from scaling when heated to high temperatures. A quick assessment of the cooling rate of an oil can also be made by the use of a nickel instead of a silver ball (Fig. 11.2). Nickel also undergoes no phase changes. In this case, the nickel ball is heated and after immersion in the oil, the time is measured (in seconds) for it to cool to its Curie point. The Curie point is the precise temperature ($354^\circ C$) at which the nickel undergoes a change in its magnetic properties. The test apparatus is designed to record the time when the nickel changes from the non-magnetic to the magnetic state. A time interval is thus obtained for the cooling process.

The second type of quench oil testing is done by carrying out actual hardening tests on quenched steel specimens. After quenching, the steel specimens are sectioned and hardness measurements made across the sectioned diameters of the steel. The variation obtained across a section is used to compare the efficiencies of different oils, with regard to their hardenability characteristics.

Industrial Oil Testing 115

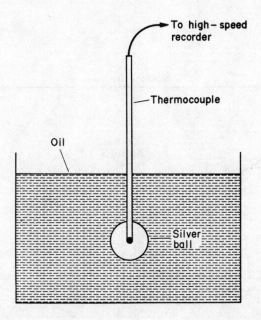

Fig. 11.1. Silver ball quench test.

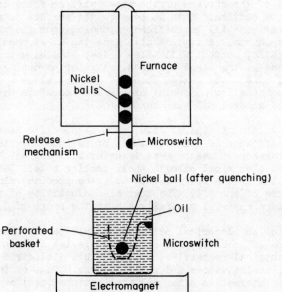

Fig. 11.2. Nickel ball quench test.

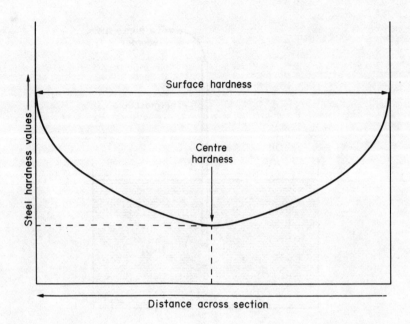

Fig. 11.3. Hardenability curve.

The hardness value of the surface of the steel will be found to be greater than that at the centre of the section. This is because the surface is cooled more quickly (Fig. 11.3). Variations will be obtained depending upon the precise steel selected to compare the quenchants. A typical oil hardening steel should be used in the test, so that the alloy content will allow a deep hardening process to occur through the entire thickness of the steel specimen. A water hardening steel containing a low alloy content, or a plain carbon steel, will not allow a deep hardening process to take place. Such shallow hardening steels are therefore unsuitable for the comparative testing of mineral oil quenchants.

Physical property values of mineral oils, such as specific heat and thermal conductivity, are of significance in heat treatment operations and other applications involving a transference of heat. Heat transfer oils are an obvious example and to a lesser extent, electrical oils in their cooling role. The specific heats of most mineral oils lie in the range 0.4 to 0.7, the precise value depends mainly upon the temperature of the oil. The thermal conductivity of an oil varies likewise with temperature and will generally lie in the range of 0.11 to 0.09 kcal/m h °C

The cooling power of an electrical oil is of significance. The electrical properties of the oil are of great importance. Many special tests have been developed to measure the electrical properties, which include dielectric strengths, electrical break-down values, resistivities and power factors. Some of these electrical properties are also of interest in other oil applications, besides their use in transformers, switch gears and cables. A good example is the cutting fluid application, which utilizes an oil as a dielectric for spark erosion or electro-discharge machining.

The ability to detect traces of water in electrical oils is of great importance. Many tests have been devised which range from a simple heating "crackle" test to much more precise elaborate procedures. The avoidance of water traces in refrigerator oils is of similar importance and accurate test procedures are again available. Special tests have also been devised to assess the solubility characteristics of refrigerator oils with critical refrigerants, over the envisaged temperature working range.

Specialised tests have been developed to assess the suitability of oils for use in the textile industry. In particular, there are tests to assess the anti-spattering or non-throw properties of oils which are used in special lubrication applications, where oil throw and possible textile cloth contamination may occur. Other tests are available to determine the scourability characteristics of oils when they are "spotted" onto cloth. The results of such tests will indicate the relative ease of removal of an oil spot by washing or scouring, if contamination should accidently occur in service. Standard colour tests are frequently carried out on mineral oils to assess their potential suitability as textile oils. Lubricant colour is also important in other applications and this is especially so for the white and technical white oil range of industrial lubricant applications.

When industrial oils are used in emulsion form, it is essential to have good stability without either water or oil separation. Large quantities of soluble oil emulsions are used, in particular, in the metal cutting industry. Standard tests have been devised to assess the emulsion stabilities of soluble oils in waters of different degrees of hardnesses. Emulsion stability is also of importance in the other applications which involve the use of the invert emulsions, such as concrete mould oil emulsions and fire-resistant hydraulic fluid emulsions.

The assessment of whether a soluble cutting oil emulsion is corrosive, or non-corrosive, to metals has usually been determined by the so-called Herbert corrosion test. Standard steel millings are placed in four small piles on a cast iron plate, the composition of which is known to be very prone to corrosion. The soluble oil emulsion is then poured to form small pools over the piles of millings in contact with the plate (Fig. 11.4). After a specified time, the plate is examined for corrosion and the assessment is made in terms of the area attacked and the intensity of the attack. A good quality soluble oil emulsion should produce no sign of corrosion. However, as with most corrosion tests, it is sometimes difficult to obtain good reproducible and repeatable results. Variations on the method have now been developed to improve the accuracy.

In the general field of fire-resistant hydraulic fluids, many tests have been devised to assess flammability characteristics. These include spray ignition tests, which involve high pressures and temperatures, autogeneous ignition tests and, for example, similar special assessment tests carried out when the fluid is mixed with coal dust. For the evaluation of fire-resistant emulsions, there are flammability tests which are carried out when various amounts of water have been allowed to evaporate from the fluid. This is to determine the loss of fire-resistance which occurs when the water content is reduced.

In the general field of hydraulic fluids, the presence of solid contaminants, such as metallic wear debris and dirt, can appreciably affect the lubrication efficiency. The size of the particles is often critical with regard to any damaging effect on the hydraulic circuit. For this reason, there are standard tests available to count not only the number of particles, which may be present in the hydraulic fluid, but also their size range. The presence of foreign particles in greases is very critical with regard to efficient lubrication in service and a standard test is also available for this particular determination.

In some applications for industrial fluids, the oil may be subjected to very high

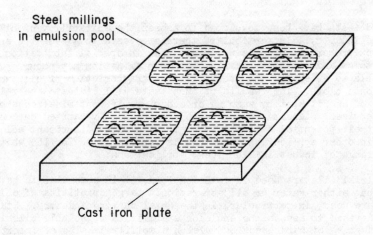

Fig. 11.4. Soluble oil corrosion test.

pressures. Examples are certain high pressure hydraulic systems and some specialised metal working operations. The viscosities of oils increase with pressure and mineral oils show greater increases than natural fatty oils. The viscosity of an oil may therefore be entirely different at its working pressure to that determined at atmospheric pressure. For this reason, pressure-viscometers have been developed to test an oil for this physical property change. One of the techniques used is to time the rate of fall of a ball, or cylinder, through the oil maintained under high pressure standardised conditions. In general, it has been found that oils which show the greatest change in viscosity with temperature, also show the greatest change with pressure.

In the general field of oil analysis, the development of spectroscopic techniques has played a very large role. These include both infra-red and ultra-violet spectroscopic test methods. These have been ably assisted by the use of the electron microscope, the mass spectrometer and X-ray diffraction methods. Another useful tool in oil analysis has been the continuing improvement in gas liquid chromatographic techniques.

The identification of minute amounts of specific hydrocarbons, in very complicated mixtures, has played an important part in the detection of possible carcinogens in oils. In the field of health and industry, many different tests are now employed and others are still being developed, to assess the possible toxicological properties of both oils and the additives in them.

It has been seen that in the general and specific testing of industrial oils, a multitude of tests are employed. Many of the general standard tests have been developed by the Institute of Petroleum (IP), in the United Kingdom and the American Society for Testing Materials (ASTM), in the United States of America. Some of the laboratory methods are joint methods which have been developed by the IP in collaboration with the ASTM. A main object has been to issue standard test techniques which are of reliable repeatability and reproducibility, when carried out in different laboratories.

In addition to these two main test bodies, there are many other authorities which have issued either suggested or specified test techniques. These include many European ones, such as the German D.I.N. committees and there are many British Standards specification test methods. There are also many defence and military specification test techniques, in addition to the very large number of industrial manufacturers' specification tests. The United States Bureau of Mines have issued many test methods, as have our own European committees concerned with safety in industry.

MECHANICAL RIG TESTING

In the last section, analytical techniques were described for the measurement of active sulphur and chlorine in cutting oils; the values obtained can be used to give an indication of the potential load carrying capacity. However, in general the testing of the load carrying capacity of industrial oils and cutting oils is best done with the use of mechanical rig test equipment. The various types of equipment use simple metal test specimens, gears or bearings. Generally, the loading is increased under dynamic conditions, until break-down of the lubricant film occurs between the test specimens. The metal to metal contact which then occurs results in scoring, scuffing or seizure of the surfaces. The load at which this happens is taken as a measure of the load carrying capacity of the oil, but in some cases frictional and wear measurements may be used in the assessment. With a background of experience of tests performed in these rigs, it is possible to provide a reliable guide to the performance of an oil under service conditions.

For the general load carrying assessment of industrial oils, the most frequently used machines are the well known Four Ball and Timken test machines. These employ metal test specimens of different geometries in rubbing contact and the test procedures are well established. For the evaluation of oils, designed for specific applications, many special machines have been developed.

Gear oils, for example, are evaluated in machines which give a simulation of gear tooth action. In the United Kingdom, the IAE gear machine, is widely accepted for the measurement of load carrying capacity. With this machine, spur test gears are used and the speed of operation for a test can be selected from a wide range. The loading is applied in defined increments to the test gears, until oil film break-down occurs which promotes surface failure by scuffing. The load at which this occurs, represents the load carrying capacity of the oil.

In Europe, the FZG gear machine is widely used for the evaluation of the load carrying ability of gear lubricants. Loading is again applied in stages and after specified time intervals, the test wheels are weighed and reassembled for a further run, with the torsional force increased by one stage. The amount of wear will be found to increase roughly by the same amount from one load stage to the next. The test run is discontinued when seizure occurs. The load carrying capacity of the gear lubricant is defined by the load stage if it is able to pass without causing seizure. The FZG test rig allows for twelve load stages and the specific wear, which is also measured, is taken into consideration in the final merit rating for the lubricant.

The assessment of the wear reduction and load carrying ability of hydraulic oils is often carried out in test rigs, which incorporate an individual manufacturer's hydraulic pump. The conditions and duration of the test are normally laid down by the manufacturer. Usually, the performance of the hydraulic oil is evaluated after wear measurements have been made of the pump components. The amount of wear must be less than a specified standard. The corrosion of the test pump components are also frequently assessed.

The evaluation of the shear stability of hydraulic oils is important when polymeric viscosity index improvers are present. The effect of shear is to cause permanent break-down of the polymer, with resultant viscosity decrease. Rigs are available to evaluate this aspect and normally they are designed to allow the hydraulic fluid under test to be pumped through a narrow orifice, in order to shear grossly the oil. Samples of the oil are drawn from the hydraulic circuit at intervals for measurement of the viscosity. The percentage decrease in viscosity and viscosity index can then be calculated. Other rigs are available which use sonic and ultra sonic methods to evaluate the tendency for viscosity index improvers to break-down under shear.

Machines are also available which can measure the friction reducing properties of lubricants, rather than their load carrying capacities. These are especially useful for example, to evaluate machine tool slideway lubricants. A measurement of the static and kinetic friction coefficients, under conditions approaching those met in practice, allows an assessment to be made of the lubricant's potential behaviour under stick-slip conditions.

The measurement of friction coefficients is useful in many other industrial oil applications. In the textile industry, laboratory friction measurements are of interest in comparing different fibre lubricants when present on the surfaces of synthetic fibres. Low friction values are required to reduce the number of fibre breakages during processing.

Coefficients of friction are also of great importance in the fields of wire drawing and metal rolling. In cold rolling, for example, the friction coefficient must be low but not too low, otherwise the steel rolls of the mill will lack grip as the sheet passes through them. Ideally, a rolling fluid should also have frictional characteristics which remain relatively constant, under a variety of physical conditions.

In addition to the frictional properties, the assessment of the reduction capacity of rolling fluids is also of importance. By reduction capacity is meant the extent a lubricant can allow the thickness of a sheet to be reduced, on a single pass through the rolls, without surface damage. In the aluminium industry, the test is carried out by coating a flat piece of aluminium strip with the test lubricant. Thi is then compressed between two steel test pieces in a special press, under carefully controlled conditions. The thickness reduction of the strip is then measured and the ability of various lubricants can be compared. In the test, the prescribed cond itions are thought to bear a resemblance to those of actual rolling in practice. These conditions are such that the aluminium undergoes severe plastic deformation by plane strain, which results in elongation of the strip.

Rig tests are essential for the extensive evaluation of greases. The equipment normally consists of ball and roller bearings which can be subjected to heavier loadings, or to temperatures more severe than those encountered in service. The tests are usually long and average times are between one and two months continuous duty.

There are many types of bearing rigs available. These include those laid down by the various manufacturers of rolling bearings. Other rigs are designed to evaluate the requirements of major customers who may, for example, use greases in electric motors. A further example of a specialised rig is an axle box lubrication rig used for the evaluation of greases for railway applications.

Despite the multitude of test rigs available for grease evaluation, they all, in the end, examine three distinct but interrelated properties of the grease. These are mechanical stability, oxidation stability and wear prevention. The performance of a grease in a test is usually first evaluated by its appearance after the test, together with the appearance and wear of the bearing elements. An indication of their interim state can often be interpreted by observation of the bearing temperatures during the test.

In the case of cutting fluids, a full assessment of an individual fluid can only be made by a controlled service trial. This is due to the many variables involved in practice; such as the different cutting speeds, types of cutting, different metals and many other factors. However, many tests are available which are made under laboratory cutting conditions and these utilize such equipment as lathes, drilling and tapping machines. These tests form a useful initial method of evaluating a cutting fluid, under carefully controlled operating conditions. The criteria rated in the tests are usually tool wear, or force measurements (such as torque) when, for instance, a tapping machine is used. The surface finish of machined components may also be used as an assessment of the cutting fluid. Dynamometers may be employed to indicate tool loadings and tool chip interfacial temperatures are also sometimes recorded. The measurement of tool wear by radioactive tracer techniques has also been done. However, it must be repeated that all these tests are only useful in a preliminary sorting capacity. A full assessment of a cutting fluid can only be made by a controlled service trial.

For the assessment of thermal stabilities of oils, which is of special importance in the case of heat transfer and heat treatment oils, various types of rigs can be designed. The tests accelerate thermal decomposition by imposing greater energy loads on the oils than they would meet in service. In the case of heat transfer oils, the oil may be circulated in a rig over electrically heated surfaces and exposed to controlled increasing quantities of heat flux. The condition of the oil is monitored by sampling at regular intervals.

In the case of heat treatment oils, the thermal stability can be assessed in a rig in which an electric resistance wire is repeatedly heated and quenched in the oil, in order to accelerate decomposition. The condition of the oil is measured after a defined quantity of electrical energy has been consumed and dissipated in the oil. Heat treatment oils may also be evaluated for stability by quenching large volumes of heated steel balls in a relatively small volume of oil, in order again to accelerate the thermal loading. It has been mentioned earlier that in the field of nuclear engineering, greases and oils used in a few specialised applications, such as remote handling devices, may be exposed to radioactive radiation. The absorption of significant amounts of radiation energy by a hydrocarbon may eventually lead to its break-down and alteration in its properties. Oils may thicken and increase in viscosity due to polymerisation reactions. Conventional soap thickened greases may first soften, due to gel structure break-down and then thicken again because of base oil hydrocarbon polymerisation. However, these properties will only change after exposure to excessive amounts of radiation. Normally, hydrocarbons can withstand moderate levels of radiation and generally the aromatic hydrocarbons are the more resistant.

Various rigs have been developed to assess the performance of lubricants, after they have been exposed to different controlled amounts of nuclear radiation. For example, bearing rig tests can be carried out on irradiated greases. The air in the atmosphere of the bearing may also sometimes be replaced by an inert gas, such as helium. As mentioned earlier, inert gas cooling is employed in certain nuclear applications and therefore it is also essential to determine the performance of a lubricant in the absence of oxygen. It is well known that normally the formation of an oxide film on a metal surface plays an important role in lubrication. The absence of an oxide film will drastically affect the type of lubricant required.

In industry, lubricants and greases are expected to function efficiently under a variety of climatic conditions. For this reason, mechanical rig tests are sometimes performed in special atmosphere chambers. These can reproduce low and high temperature environments, under a variety of conditions. For example, rig tests can be carried out in both dry and wet atmospheres. If the particular piece of industrial equipment to be lubricated is relatively small and portable, then a full-scale test can be carried out in the chamber under controlled laboratory conditions. After the test, the equipment can be stripped down and the appropriate wear, deposit or other measurements made under ideal surroundings.

In the case of industrial engine oils, the only accepted criteria for quality are the results of engine tests carried out using the oils. Standardised test engines have been developed to classify oils for use in compression ignition engines. The main test criteria utilized in such tests are the cleanliness of the engines, together with wear measurements.

Field Testing

With large pieces of plant and equipment, the mechanical rig test procedure, under controlled laboratory conditions, is not practicable. The only way then is the arrangement of a full-scale field trial, after a limited amount of initial sorting laboratory testing has been carried out. When conducting a field trial of any description, it is first essential to ensure that the plant or equipment is thoroughly cleaned before the test oil is introduced. Any other precautions which can be taken to reduce the number of variables in the trial should also be carried out. The trial must be controlled as carefully as possible. It may be necessary to take oil samples at intervals or make wear measurements of equipment parts, or assess the condition of the components being produced. The type of measurement necessary will vary depending upon the type of industry and trial.

The test criteria to be used must be clearly established. For example, the trial may be a purely comparative one, in which a lubricant is being tested against an alternative lubricant. The trial may involve the continuous running of a piece of equipment for a considerable length of time, without break-down or service trouble. Alternatively, the trial may be the attainment of a certain component production figure, without lubrication failure. In certain instances, the quality of the surface finish of the components being produced may be the most important criteria.

In large works, such as encountered in the steel industry, it is extremely difficult to arrange for the full-scale testing of a lubricant or rolling fluid. This is due to the possible interference with production which may occur in the event of a lubricant failure, which would result in a considerable cost. In addition to direct production losses, the replacement of damaged plant due to the lubrication failure, may not only be extremely costly but also time consuming with further indirect losses of production. However, despite the difficulty sometimes encountered in arranging a field trial, it is the only way to evaluate correctly a lubricant under practical conditions. The testing of new experimental lubricants under practical conditions leads the way to the possible discovery of improved production rates or machinery performance. Even if such practical improvements are not obtained, a trial may still reveal to a lubrication engineer an alternative supplier for a satisfactory product, or an alternative product which may have a price advantage. . .

CHAPTER 12

Test Terms and Standard Methods

Oil Test Terms

Aniline point. The aniline point of an oil is the lowest temperature at which it remains completely soluble with an equal volume of aniline. The test gives an indication of the aromaticity of the oil.

Ash content. The ash content of an oil is the weight percentage remaining after the incineration of the oil has been carried out under prescribed conditions.

Asphaltene content. The weight percentage of wax free material in an oil, which is insoluble in normal heptane but soluble in benzene.

Auto-ignition temperature. The temperature at which an oil will spontaneously ignite, when heated without the application of a naked test flame or spark. It can also be defined as the temperature at which ignition of oil will occur by contact with a hot body.

Carbon residue. Carbon residue values of petroleum products are expressed in weight percentage terms, after the controlled volatilisation of the product has been carried out and the residue weighed. The results will normally be quoted as either Conradson or Ramsbottom, depending upon the test method used.

Cloud point. The cloud point of an oil is the temperature at which paraffin wax starts to separate, or crystallise, from the oil when it is cooled under prescribed conditions.

Demulsification value. This measurement expresses the time (in seconds) required for water separation to occur from an oil and water mixture, which has been prepared under standardised conditions.

Drop point. This test evaluates greases. The drop point of a grease is the transition temperature at which the semi-solid state changes to a liquid state. The drop point is determined by heating the grease under controlled conditions.

Fire point. The fire point is the temperature at which an oil starts to burn continuously, when heated under prescribed conditions in the presence of a flame. The test is carried out in the same apparatus as employed for the determination of the flash point (open) and is a continuation of this test. As a rough estimate, the fire point of a mineral oil will normally be approximately $40°F$ to $60°F$ above the open flash point.

Flash point (closed). The temperature at which the vapour from an oil will ignite, in a closed test apparatus of stipulated design and dimensions. A flame is used as the ignition source and the oil is heated under controlled conditions.

Flash point (open). The temperature at which the vapour from an oil will ignite, when heated under controlled conditions in the presence of a flame, in an open test apparatus of stipulated design and dimensions. The open flash points of most mineral oils will, as a rough guide, be in the region of $20°F$ to $30°F$ above the closed flash points.

Iodine value and bromine number. The iodine value is the quantity of iodine, expressed in grams, which reacts with 100 grams of the oil in a standard test. Iodine values give an indication of the amount of unsaturated hydrocarbons present in an oil. A similar test to assess the content of unsaturated materials is the bromine number. Bromine replaces the iodine as the test reagent. The bromine number test is particularly used for the assessment of aluminium cold rolling fluids.

Neutralization value. This value denotes the amount of alkali required to neutralize the acidic products in a specified weight of oil. The neutralization value is normally expressed as the number of milligrams of potassium hydroxide required to neutralize one gram of oil. The acidic products are usually organic in nature. Mineral or inorganic acids are normally absent from oils unless contamination has taken place.

Penetration value. A test normally reserved for measuring the softness or consistency of greases. The test is carried out with a penetrometer. This instrument allows a standard cone to sink freely into a sample of the grease for five seconds, at a temperature of $25°C$. Before the test, the grease is worked by agitation. The depth of penetration is measured in tenths of a millimetre and the result is quoted as the worked penetration value for the grease.

Pour point. This denotes the lowest temperature at which an oil will just flow, when it is cooled under prescribed test conditions.

Saponification value. The saponification value of an oil is the number of milligram of potassium hydroxide which are required to saponify one gram of the oil. It is frequently used to indicate the presence of fatty materials in oils. A saponification reaction involves the formation of a soap, together with glycerol, when a fatty material is broken down by alkali reaction. With a simple mineral and fatty oil blend, the saponification value can be used to calculate the amount of saponifiable matter present in the oil.

Specific gravity. The specific gravity of an oil may be defined as the ratio of the mass of a given volume of the oil at a temperature, t_1, to the mass of an equal volume of pure water at a temperature, t_2. In the petroleum industry, specific gravities are frequently quoted with the two temperatures equal. However, the temperatures t_1 and t_2 are normally always reported. An example of the designation employed is specific gravity $60°F / 60°F$ and this is followed by the precise gravity figure.

In the United States of America, an arbitrary gravity figure is often used, known as the A.P.I. gravity, which is quoted in degrees. The initials denote the American

Petroleum Institute. The A.P.I. gravity is related to the specific gravity 60°F / 60°F (S). Both gravity figures can be inter-converted by the use of the following equation

$$°A.P.I. = \frac{141.5}{S} - 131.5$$

$$\text{or} \quad S = \frac{141.5}{131.5 + °A.P.I.}$$

Specific heat. The specific heat of an oil can be roughly calculated from its specific gravity value at 60°F / 60°F. The calculation may be made with the assistance of the following equation

$$\text{Specific heat} = \frac{0.403 + 0.00081\ T}{\text{Square root of specific gravity}}$$

T denotes the temperature of the oil in degrees centigrade. The specific heat of an oil therefore varies with temperature. A typical mineral oil possesses a specific heat of approximately 0.45 at a temperature of 60°F (15°C).

Sulphur content. The sulphur content of an oil normally denotes the total sulphur present in the oil and is expressed in weight percentage terms. However, the total sulphur figure draws no distinction between sulphur which is completely unreactive and that which is corrosive to metals. When the distinction is important, corrosive sulphur is qualitatively detected by heating the oil in the presence of a copper strip and observing whether discolouration of the copper occurs. When a quantitative assessment of corrosive sulphur is required, as for example in a cutting oil, then a more elaborate test method is used. This involves the actual measurement of the quantity of sulphur in the oil which is reactive to copper. The test is normally carried out at two different temperature levels.

Thermal conductivity. The thermal conductivity of an oil varies with temperature (T). It can be roughly calculated when the specific gravity value of the oil at 60°F / 60°F is known. This is done with the use of the following equation

$$\text{Thermal conductivity (kcal/m h °C)} = \frac{0.101}{\text{Specific gravity}} \times (1 - 0.00054\ T)$$

A typical mineral oil will possess an approximate conductivity value of 0.11 kcal/m h °C at 60°F (15°C).

Viscosity. The viscosity of an oil can be defined as the ratio of the shearing stress to the rate of shear. At a particular temperature, this ratio is normally constant with most mineral lubricating oils; in other words, the rate of shear is proportional to the shearing stress. Such oils are said to be Newtonian in character. In simple terms, the viscosity of an oil may be viewed as its resistance to flow or its own internal friction. A thick oil has a high viscosity, whilst a thin one has a low viscosity.

Dynamic viscosity values are measured in units of a poise. For practical purposes, the centipoise is more frequently used. The dynamic viscosity can be simply defined as the force necessary to maintain a 1 square centimetre plate moving at a constant velocity of 1 centimetre per second, when separated from a stationary parallel plate by a distance of 1 centimetre. The space between the plates is filled with the oil. As the plate moves, the oil will also move or flow under streamline conditions. If the force needed is one dyne then the viscosity of the oil is one poise.

However, in the petroleum industry, the viscosity of an oil is expressed in various units but rarely in dynamic units. The most common units are Stokes or centistokes, Redwood Seconds, Engler degrees and Saybolt Universal Seconds. The various viscosities can be approximately converted from one to another at the same temperature level, by the use of a standard conversion table (see appendix, Table 1).

In the actual measurement of viscosity in the petroleum industry, various viscometers are used depending upon the type of unit measurement being carried out. Most methods rely on the measurement of the time taken for a known volume of oil to flow through a standard orifice or tube. It will be apparent that in these types of viscometers, the oil flow through the tube is initiated by the force of gravity. The temperature grossly affects the viscosity of the oil, so this variable must be closely controlled in the test. The exact test temperature is therefore always measured and quoted alongside the viscosity value.

There is a relationship between the kinematic viscosity (cSt) and the dynamic viscosity (cP), when the density of the oil at the test temperature is known. The kinematic viscosity value is equal to the dynamic viscosity value divided by the oil density value.

Viscosity index. The viscosity index of an oil is an empirical figure which denotes the effect of temperature change on the viscosity of the oil. The higher the viscosity index level, then the smaller the expected viscosity change. The viscosity index of an oil is determined from kinematic viscosity measurements made at two temperature levels, namely $100°F$ and $210°F$. The determined kinematic viscosity values are them employed to calculate the viscosity index of the oil.

However, before this calculation can be carried out, two other theoretical figures are required which are selected by consulting standard tables, issued by the Institute of Petroleum in the United Kingdom. These tables are used to find the viscosity values at $100°F$ of a 0 viscosity index reference oil and also a 100 viscosity index reference oil. Both reference oils must possess the same viscosity value as the test oil at $210°F$. The equation employed for the calculation of the viscosity index is as follows

$$\text{Viscosity index} = \frac{100 (L - U)}{L - H}$$

U is the measurement value obtained for the kinematic viscosity of the test oil in centistokes at $100°F$.

L is the viscosity at $100°F$ of an oil of 0 viscosity index, having the same viscosity in centistokes as the test oil at $210°F$.

H is the viscosity at $100°F$ of an oil of 100 viscosity index, having the same viscosity in centistokes as the test oil at $210°F$.

In the standard tables, it will be found there are three columns arranged side by side. The first column lists kinematic viscosity figures at $210°F$, the second column lists the equivalent L values and the third column lists the equivalent L - H values. The appropriate L and L - H values, which correspond with the measured kinematic viscosity value for the test oil at $210°F$, are easily selected from the tables.

The viscosity index calculation method is applicable to the majority of the oils used as industrial lubricants. The exceptions are oils of extremely low, or high, viscosities and those with relatively high cloud points. It is worth noting that a viscosity index figure of 100 was originally selected to represent the viscosity of a Pennsylvanian paraffin base oil of good quality. A viscosity index of 0 represented an oil of poor variable quality. However, for many years now it has been possible to refine mineral oils to possess viscosity indices in excess of 100, without the use of additives. The incorporation of viscosity index improvers yields oils of even higher values.

Standard Test Methods

The precise laboratory procedure details of many of the various tests described in the book can be obtained by consulting a standard reference book, such as that issued by the Institute of Petroleum, in the United Kingdom or the American Society for Testing Materials in the United States of America.

The "IP Standards for Petroleum and its Products. Part 1 - Methods for Analysis and Testing" contains laboratory test procedures for all petroleum products and it also includes a few small scale rig tests. Most of the methods are recognized as standard methods but some are tentative and proposed methods of test. Some of the methods are joint with the American Society for Testing Materials (ASTM). The following alphabetical list gives the particular methods which may be of interest in the field of industrial lubrication and reference to them in the IP book will give full experimental details of each individual test

Acidity and alkalinity of greases
Acidity (inorganic) of petroleum products - colour indicator titration method
Acidity of petroleum products - neutralization value
Adhesion and stickiness - hard-film temporary protectives
Air release value
Analysis of oil-soluble sodium petroleum sulphonates
Aniline point
Anti-wear properties of hydraulic fluids - vane pump test
Aqueous cutting fluid - corrosion of cast iron
Ash from grease
Ash from petroleum products
Ash from petroleum products containing mineral matter
Asphaltenes - precipitation with normal heptane
ASTM colour of petroleum products - ASTM colour scale

Barium, calcium, magnesium and zinc in unused lubricating oils by atomic absorption spectroscopy
Barium, calcium, phosphorus and zinc in lubricating oils - direct reading emission spectrographic method
Barium in lubricating oil
Barium in lubricating oil additives
Bromine index of petroleum hydrocarbons by electrometric titration
Bromine number - colour-indicator method

Calcium in lubricating oil
Calculating viscosity index from kinematic viscosity
Chlorine in new and used lubricants (sodium alcoholate method)
Chlorine in petroleum products. Flask combustion method
Cloud point of petroleum oils
Colour by the Lovibond tintometer
Cone penetration of lubricating grease
Cone penetration of lubricating grease using one-quarter and one-half scale cone equipment
Cone penetration of petrolatum
Cone penetration of petrolatum - rapid method (proposed)
Congealing point of petroleum waxes, including petrolatum
Conradson carbon residue of petroleum products

Corrosive substances in grease - copper strip test
Corrosive sulphur in electrical insulating oils
Counting and sizing particulate matter in aerospace hydraulic fluids using a HIAC particle counter

Demulsification number - lubricating oil
Density and relative density of liquids by Bingham pyknometer method
Density at 20°C
Density, specific gravity, or API gravity of crude petroleum and liquid petroleum products (hydrometer method)
Detection of copper corrosion from petroleum products by the copper strip tarnish test
Distillation of crude residue
Distillation of petroleum products
Drop melting point of petroleum wax, including petrolatum
Drop point of grease
Dropping point of lubricating grease
Drying characteristics - hard-film temporary protectives
Dynamic anti-rust test. Lubricating greases

Electric strength of insulating oils
Electric strength - transformer oils
Emulsifiable cutting oils. Emulsion stability
Emulsion stability of water-in-oil emulsions at ambient temperatures
Engine cleanliness - Caterpillar high speed supercharged compression-ignition test engine
Engine cleanliness - Petter AVI compression-ignition test engine
Engine cleanliness - Petter AVB supercharged compression-ignition test engine
Evaporation loss of lubricating greases
Extreme pressure properties, anti-wear, friction and corrosion protection properties of fluid lubricants: Falex machine
Extreme pressure properties, friction and wear for lubricants: four-ball machine
Extreme pressure properties, friction and wear for greases: Timken wear and lubricating testing machine
Extreme pressure properties, friction and wear for lubricating fluids: Timken wear and lubricant testing machine

Ferrous corrosion - prevention characteristics of gear oils in the presence of water
Flash point by Pensky-Martens closed tester
Flash point (open) and fire point by means of the Pensky-Martens apparatus
Foaming characteristics of lubricating oils
Foreign particles in greases
Frothing characteristics of emulsifiable cutting oils

Heat-stability of calcium-base greases
Hydrocarbon types in liquid petroleum products by fluorescent indicator adsorption

Iodine value - iodine monochloride method

Kinematic viscosity of transparent and opaque liquids and the calculation of dynamic viscosity

Lead, copper and iron in lubricating oils
Lithium and sodium in lubricating greases by means of a flame photometer
Load-carrying capacity tests for oils - IAE gear machine
Low-temperature torque test - lubricating greases

Measuring frictional properties of slideway lubricants
Melting point of petroleum wax (cooling curve)

Microscopically sizing and counting particles from petroleum-base aerospace
  hydraulic fluids

Neutralization number by colour-indicator titration
Neutralization number by potentiometric titration
Non-normal paraffins content of fully refined petroleum waxes (proposed)

Odour of petroleum wax
Oil content of petroleum waxes
Oil separation on storage of grease
Olefins and aromatics - calculation
Oxidation characteristics of inhibited steam-turbine oils
Oxidation stability of inhibited mineral turbine oils
Oxidation stability of lubricating greases by the oxygen bomb method
Oxidation stability of mineral insulating oil
Oxidation stability of steam turbine oils by rotating bomb
Oxidation stability of straight mineral oil
Oxidation test for transformer oil

Penetration of bituminous materials
Penetration of semi-fluid greases
Phosphorus in lubricating oil - quinoline phosphomolybdate method
Pitting failure tests for lubricating oils in a modified four-ball machine
Pour point of petroleum oils

Ramsbottom carbon residue of petroleum products
Rapid tests for flash point
Refrigerator oil - dichlorodifluoromethane insoluble content
Relative density and density
Rolling bearing greases. Churning test
Rolling bearing performance test - lubricating greases
Rust prevention characteristics of soluble oil emulsions - chip/filter paper method
Rust preventing characteristics of steam-turbine oil in the presence of water
Rust preventing characteristics - temporary corrosion preventive

Sampling petroleum and products - liquids, semi-solids and solids
Saponifiable and unsaponifiable matter - oils, fats and waxes
Saponifiable number of petroleum products
Seal compatibility index of petroleum products
Shear stability of polymer-containing oils using a diesel injector rig
Soluble cutting oil - oil content of dispersions
Specific refractivity
Spectrographic analysis - inorganic constituents of ashes
Sulphated ash from lubricating oils and additives
Sulphur in petroleum products
Sulphur (reactive to copper) in cutting oils

Temporary corrosion preventives - water displacing and corrosion protection properties
The assessment of lubricants by measurement of their effect on the rolling fatigue
  resistance of bearing steel using the unisteel machine
The assessment of the oxidation stability of mineral turbine oils during use
Thermal stability of emulsifiable cutting oil
Total base number of petroleum products by potentiometric perchloric acid titration
Total solids in used engine oils

Vapour pressure - micro method

Water and sediment - centrifuge method
Water washout characteristics of lubricating greases

Zinc in lubricating oil

It should be noted that the selected methods listed are reviewed on a continuous basis by the Institute of Petroleum and revised editions of the standard reference book are issued on a regular basis. It is therefore advisable to employ the latest edition when searching for the details of one of the individual general laboratory methods. Similar comments apply when the American Society for Testing Materials standard reference book is employed.

APPENDIX

TABLE 1 Viscosity Conversion (Approximate)

Centistokes (cSt)	Saybolt Universal Seconds (SUS)	Redwood 1 Seconds	Engler Degrees
5	42	39	1.4
9.1	56.5	50	1.74
14.1	75	65	2.3
17	86	75	2.51
20.2	100	88	3.0
24	115	100	3.35
25	119	105	3.45
30	141	125	4.1
36	168	150	4.85
41.8	198	170	5.5
45	209	185	6.0
49	230	200	6.5
54.3	250	212	7.0
55	255	226	7.3
62	288	250	8.1
65	300	266	8.6
70	324	287	9.25
74	346	300	9.7
80	371	328	10.55
86	400	350	11.4
98	460	400	13.0
100	463	410	13.2
110	500	443	14.5
124	575	500	16.2
130	600	533	17.15
140	649	574	18.5
148	690	600	19.4
150	700	615	19.8
160	742	656	21.1
172	800	708	22.8
185	860	750	24
194	900	795	25.6
200	927	820	26.4
216	1000	886	28.5
220	1020	990	29.05
245	1300	1000	32

Bibliography

Bradbury, F. (1972) <u>Hydraulic Systems and Maintenance</u>, Iliffe, London.
Braithwaite, E.R. (1967) <u>Lubrication and Lubricants</u>, Elsevier.
Braithwaite, E.R. (1963) <u>Solid Lubricants and Surfaces</u>, Pergamon Press, Oxford.
Evans, E.A. (1963) <u>Lubricating and Allied Oils</u>, Chapman and Hall, London.
Houghton, P.S. (1970) <u>Gears</u>, Technical Press, London.
Lissaman, A.J. and Martin, S.J. (1966) <u>Principles of Engineering Production</u>, English Universities Press, London.
Neal, M.J. (1963) <u>Tribology Handbook</u>, Butterworth, London.

Index

Abrasive environments 91
Acidity value 111, 113
Additives 5-15
  anti-foam 8, 22
  anti-wear 29
  detergent dispersant 10-1, 93
  fatty 9, 13-5, 30, 37, 94
  friction improver 9, 37, 62
  load-carrying (extreme pressure) 8, 36, 63, 103
  oxidation inhibitor 6-7, 77, 86, 113
  pour point depressant 9, 77, 88
  rust (corrosion) inhibitor 7-8, 90
  surface active 11-2, 42, 82, 96, 104
  viscosity index improver 10, 120
Air compressors 30
Air filter oils 94-5, 98
Air release 28-9
Air tool oils 37
Aluminium 62, 65, 71, 98
American Society for Testing Materials 119, 127, 130
Aniline point 112, 123
Anti-foam 8, 22, 28-9
Anti-seize 24
A.P.I. gravity 124-5
Aromatics 5-6, 22, 74, 88-9, 97, 100
Ash content 112, 123
Asphaltenes 112, 123
Auto-ignition 111, 123

Bacteria 53, 57, 61-2, 98, 104, 106, 114
Bearings
  air 73
  ball and roller 18, 47, 51, 55
  heat transfer effects 72-3
  lubricants 26-7, 39-40
  plain 18, 87
Biocides 62, 105
Bitumen 45, 56, 100-1
Blow holes (concrete) 96

Boundary lubrication conditions 2-3, 68, 86
Brick manufacturing industry 4
Bright oils 94
British Standards 119
  viscosity grades 4-5
Broaching 62-3
Bromine number 113, 124
Built-up edge 59

Cable oils 78-9
Carbon
  black 100
  deposits 30
  residue 112, 123
Carcinogens 64, 99, 118
Care of industrial oils 102-10
Cast iron 71
Castor oil 14
Cemented carbide tools 63
Cement manufacturing industry 4, 91
Centralised systems 17, 27, 31, 49, 51
Centrifuging 108
Ceramic tools 63
Chlorinated hydrocarbons 19-20, 63, 74-5, 77
Chlorine 8, 63, 114
Circulation oils 25-34
Cleaning (metal) 58
Cloud point 10, 112, 123
Coal mining industry 3
Cold extrusion 68-9
Compressors 30, 38-9
Concrete mould oils 95-7
Coning oil 42, 99
Conradson 112, 123
Consistency (grease) 49-50, 113, 124
Coolers 84, 87
Copper 62-3, 71
Corrosion preventives 51-8
Corrosive environments 90, 113
Crackle test 117

Cutting fluids 1, 59-65, 71, 114, 121

Dark cylinder oil 94
Deep drawing 69
Demulsification 11-2, 123
Dewatering 53-4
Dielectric 78
Diesel 11, 92-4
Diesters 20-1, 87
Diphenyl-diphenyl oxide 74
Disposal 62, 105
Drag-out 64, 84
Drip feed 18
Drop point 114, 123

Efflorescence (concrete) 96
Electrical oils 77-9, 102, 116-7
Electro-discharge machining 69-71
Emulsifiers 11-3, 62
Emulsions 12-3, 37, 41, 60, 96, 104-6, 117
Engine oils 92-4, 122
Eutectic salts 74
Expansion tank 73, 76
Explosions 30, 105

Fatty oils 9, 13-5, 30, 62, 86
Field testing 122
Filtration 64, 66-7, 84, 91, 94-5, 103-4, 107
Fire 30, 83-4, 87, 105
    point 111, 123
    resistant lubricants 19-20, 57, 87
Flash point 111, 124
Flushing 77, 104
Food industry 4, 97-8, 100
Force feed 18, 93
Four ball test 119
Free-scouring oils 38
Friction
    coefficients 37, 120
    reduction 36
FZG gear machine 119

Gears 31-4
    cutting 62
    fretting 90
    FZG 119
    IAE 119
Gelling agents 13, 95
Glycol 21, 42, 82, 106
Graphite 23-4, 49, 69
Greases 16, 47-51, 87, 113, 120-1
    aluminium 34, 47
    calcium 47, 55

clay 47
complex 47
lithium (multi-purpose) 47
nuclear 88-9
polymer 34, 48
silicone 21, 47
sodium 47

Hand oiling 18, 45
Hardenability 114, 116
Health (hygiene) 41, 64-5, 97, 99, 104, 107
Heating 102
Heat transfer oils 73-7, 98, 121
Heat treatment oils 79-85, 121
Herbert corrosion test 117-8
High temperature environments 86-7
Hydraulic oils 27-9, 117, 119-20
    fire-resistant 20, 29, 105, 117
Hydrodynamic lubrication 1-2, 9, 35, 68

IAE gear machine 119
Institute of Petroleum 119, 126-30
Iodine value 112-3, 124

Jute 41

Kerosine 69
Knitting 42

Laminar flow 75
Lanolin 53
Lead 8
Load carrying 8, 119
Low temperature environments 87-8
Lubricant application 16-9
Lubrication survey 17-9
Lubricators 16, 18, 45, 93

Machine tools 17, 36, 61-2
Magnesium 71
Manufacturers' recommendations 16
Marquenching 82
Martensite 79-80, 82
Mechanical rig tests 119-22
Mechancial stability 49
Metal working fluids 65-71
Micro-fog 18
Mill-scale 53
Mineral oil (types) 4-6
Mist 31, 42, 103-5
Molybdenum disulphide 23-4, 49
Mould release 96

Multi-stand mill 26

Naphthenes 5-6
Naphthenic oils 9, 39, 77
National Lubricating Grease Institute 49-50
Neutralization value 111, 124
Nickel ball test 114-5
Non-ionic 12-3
Non-throw oils 38
Nuclear radiation environments 88-9, 121-2

Oleines 40
Olive oil 40
Oxidation stability 6-7, 49, 77, 86, 113
Oxygen compressors 30
Ozone 107

Palm oil 15, 65
Paper manufacturing industry 3
Paraffins 5-6
Penetration (grease) 113-4, 124
Perfluorocarbons 20
Petrolatum 53, 55-6, 99-100
Petroleum
  extracts 100
  jelly 45, 90, 99
  sulphonate 12, 97
Pharmacopoeia 97
Phosphate esters 19-20, 87
Phosphating 68
Phosphorus 8, 63
Planned lubrication 19
Polynuclear aromatics 64
Polytetrafluoroethylene 18, 20, 36
Pour point 9, 39, 112, 124
Power factor 78
Pressure viscometer 118
Production oils 1, 59
Pumpability (grease) 49

Quenching oils 79-85, 114

Ramsbottom 112, 123
Reclaimation (recovery) 64, 107-8
Refining 4, 10, 65, 78, 99-100
Refrigerator oils 38-9
Retarded hydration (concrete) 96
Ring oiling 18, 39, 93
Rock drill oils 37
Roller trough 42

Rolling
  aluminium 39-40, 65-7, 98, 120
  steel 26-7, 57, 65-6, 87
Rubber industry 100
Rusting 7, 51-3, 113

Safety 65, 84, 105
Sampling 108-10
Saponification value 111, 124
Screwing 62
Shear 48, 114, 120, 125
Silicates 74, 87-8
Silicones 21, 89
Silver ball test 114-5
Skin irritation 62, 65, 104
Slack waxes 100
Slideways 35-7, 120
Solid lubricants 2, 22-4, 89
Space environments 89-90
Spark erosion 69-71
Specific
  gravity 112, 124
  heat 116, 125
Spin finish 42
Standard test methods 127-30
Staple (yarn) 42-3
Steam 73, 103
  engines 94
  turbine oils 25-6, 113
Steel industry 3, 17, 51, 122
Steels 53, 71, 79-85, 90, 116
Stick-slip 36-7
Storage 53, 91, 102
Sulpho-chlorinated oil 63
Sulphur 8-9, 63, 90, 93, 112, 114, 125
Sulphurised fat 63
Swarf 63-4
Switch oil 77
Synthetic oils 2, 19-22, 33, 39-40, 42, 87, 107

Tapping 121
Tempering 82-3
Temporary corrosion preventives 51-8
Terphenyls 74
Testing 123-30
Textile lubricants
  fibre 40-3, 99, 120
  machinery 3, 37-8, 99, 117
Texturising 41-2
Thermal conductivity 116, 125
Throw-away tips 63
Timken test 119
Titanium alloys 71
Transformer oils 77
Triaryl dimethanes 74

Turbine oils 25-6, 113
Turbulent flow 75
Twort 99

Undersealant (chassis) 101
United States Bureau of Mines 119

Vacuum
　environments 89-90
　pumps 30-1
　quenching 82, 84
Vapour pressure 30-1, 89
Viscosity 4, 9, 17, 29, 49, 72, 125-6
　index 9-10, 29, 88, 126-7

Wax 39, 42, 88, 100
White oils 97-9
Wire
　drawing 67-8
　rope lubricants 43-6
Wool 40-1
Work hardening 71
Worsted 40

Yield value 49